Villa Maior

The Viniculture of Claret

A Treatise on the Making, Maturing and Keeping of Claret Wines

Villa Maior

The Viniculture of Claret
A Treatise on the Making, Maturing and Keeping of Claret Wines

ISBN/EAN: 9783743403772

Manufactured in Europe, USA, Canada, Australia, Japa

Cover: Foto ©berggeist007 / pixelio.de

Manufactured and distributed by brebook publishing software (www.brebook.com)

Villa Maior

The Viniculture of Claret

THE VINICULTURE

OF

CLARET

A TREATISE ON THE MAKING, MATURING, AND KEEPING OF CLARET WINES

BY

THE VISCOUNT VILLA MAIOR

Translated by REV. JOHN I. BLEASDALE, D. D

Organic Analyst and Œnologist, &c

SAN FRANCISCO, CAL

PAYOT, UPHAM & CO., PUBLISHERS

1884

Printed by Geo. Spaulding & Co.
414 Clay Street.

SAN FRANCISCO, Cal., 23d June, 1884.

ARPAD HARASZTHY, ESQUIRE, President of the
Vinicultural Society and California State
Board of Viticultural Commissioners:

Dear Sir—I have much pleasure in dedi-
cating to you the translation of this, treatise
by Viscount Villa Maior, of Portugal, because
at the present time, more than ever before,
accurate and minute information concerning
absolutely all the details of the making, treating and keeping of CLARET is required, if our
vintners are to compete successfully with the
higher grade of French red wines, which will
alone in future be sent into American markets;
and because of your untiring life-labors to
place the wines (especially the Clarets) of California in the high position they are capable
of holding among the wines of the world; and
lastly, because a seven years' residence in
Portugal, and a five years' residence here,
enable me to speak with confidence of their
similarity in both climate and soil, and the
adaptability of the practical directions of this
small treatise to meet present requirements.

Faithfully yours,

JOHN I. BLEASDALE,
Organic Chemist and Œnologist.

TO THE READER.

This treatise, in twelve short chapters, is here offered to those interested in Claret Wines. It hardly touches on white wine, the object of the author—himself one of the most renowned practical wine men of the world—being to teach his ignorant countrymen how to make a wine capable of competing successfully in the English markets, not more than 26 British proof, so that it might enter English marts under the same tariff as French red wine.

J. I. BLEASDALE.

INTRODUCTION.

The makers of wine are numerous enough, but few make it well. Yet to make wine well, is not a matter so difficult as to exceed the comprehension of ordinary understandings.

Beyond all doubt, there exist in Portugal all the requisite conditions for the producing of numerous distinct, and very various kinds of wines, the major part of which are capable of acquiring excellent qualities, and of constituting a most important branch of commerce, and a great increase of public wealth. But it is also certain that beyond the generous wines of the Douro and Madeira, those of the Bairrada and some of the Estremadura, the rest have but a very poor show for exportation; while yet they might and ought to hold a preponderance in foreign markets, seeing that the principal wine-using countries demand at the present day, principally genuine wines, moderately alcoholic, nourishing, and fit for ordinary drinking.

Our export business in this branch has not yet begun; yet it is just from it that we ought to derive our greatest wealth; because it is just this class of wines which we can produce on the greatest scale, and at least cost, thus being in a position to supply them cheap, circumstances conducible to the extension of the wine industry and the increase of profit.

In regard to this kind of wines, we have hitherto contented ourselves, for the most part, with supplying our domestic requirements from year to year, to the extent that at the arrival of next year's vintage hardly any remains from the preceding year, and that itself so badly made that but little of it reaches the end of the summer without having commenced to show signs of decay. On this account, probably, it has not found its way into commerce for exportation.

It certainly is not due to the natural conditions of our viticultural regions; nor, generally speaking, to the bad quality of the grapes employed in making our wines of consumption, to which we can attribute the imperfections we find in them; but to original defects in the making, and the little or no care bestowed on their keeping.

With the view of rendering some assistance towards the improvement of our pure nutritious wines, whether for home consumption or export, I am writing this compendium of the principal rules, precepts, and principles governing the management, etc., of wines; as being specially adapted to the conditions of our different wine regions.

GENERAL CONSIDERATIONS.

Although there may be innumerable varieties of wines, between the commonest and the highest class, still a *well made wine* is always a *good* wine, within the limits of its category.

The qualities which any given wine should have, to entitle it to be considered good, in its class, stand always in relation to the nature of that wine. Thus the qualities of a good green *(vinho verde)* wine, no matter from what quarter of Minho, ought not to be compared with the qualities of good ripe wines of other parts of the country; and for the same reason, in order that we should consider good a wine of ordinary consumption, we should not compare it with generous wines of a superior class. Absolute equality between things, as well as between men, is a chimera.

For each species of wine there ought to be a perfect type to which all of that class should be referred. This typical wine ought to be the product of the very best quality of grapes grown in the district, and which have attained the most perfect maturity; made with all care; free from the impurities which might endanger its keeping qualities, and finally nursed up to the highest point of perfection of which it is susceptible.

All *good* wines, in the order to which they belong, are pleasant to the palate, on account of their peculiar relish and perfume; to the sense of smell by their bouquet, and to the sight by their color and brilliancy which attest their purity. All should be easily digested, in order to be healthy, neither prejudicing the health nor the reason, when used in moderation, relatively to their nature.

Many are the kinds and varieties of wines which we can make. It would be in the highest degree advantageous, if we had a perfect and definite classification and differentiation of them all; but since all the elements indispensable for it are not at present in existence, we can, at least provisionally, content ourselves with distributing in the following groups, viz:

1.   Ordinary and fine table wines, both for home consumption and for export.
2.   Generous and alcoholic wines of great endurance.
3.   Liqueur and special wines.
4.   Wines fit only for distillation.

Every one knows that the recognized species and varieties of wines depend for their nature and qualities on the grapes they are made from, and the methods followed in the making.

The vineyardist who undertakes to direct the making of his own wine, before all, should have a clear idea of the exact result he seeks to obtain; that is to say, of the character of wine he hopes to obtain, so as to employ the material and method best calculated to secure the object he is aiming at. To make wine or anything else at haphazard and empyrically, is to act irrationally. And so likewise he should always have before his mind the type best suited to the nature of his produce, and to know thoroughly the process by which the typical wine was made.

In a country like ours, where the conditions for making wine are so grand and so various, it is almost impossible to lay down fixed and invariable rules for making all the different kinds of wine, which may be rigorously applied in the several districts. It is out of the question, in a small work like the present, that I should descend to all particulars which are interesting to the different wine districts. I shall confine myself to a description of general methods, applicable to the great majority of cases; and chiefly to the manufacture of generous and nutritious wines, which are comprised in the first group as stated above; and are the kinds which can have and ought to have increased consumption, and produce a more extensive and lucrative export.

In the making of all kinds of wines, there are certain general rules which it is well to have at one's finger-ends, applicable to all cases; and which, once fully comprehended, can be readily modified and applied in especial cases, which are determined by the condition of particular localities. I will endeavor to expound these rules, and show their *raison d'etre*, indicating, as far as possible, the variations and exceptions necessary to be attended to in the more remarkable of particular cases.

THE VINTAGE.

Wine-making is the ultimate aim and object of viticulture, the multifarious operations of which ought to be directed to the producing of good grapes, such as alone yield good wine. I have no intention at present of entering upon vine-yard cultivation, a subject large enough for an entire treatise; and so I will leave it at the point when the grapes arrive at a state of ripeness fitting them to make wine, for here a true industry commences in which the primary material is the grape, and the product — wine. Upon the vintage, which comprises the gathering and preparation of the grapes, as the primary material of wine, depends in a great degree the result of the wine making, which is the essential operation of this industry.

The prudent vineyardist, before commencing to gather his grapes, should bear in mind the various conditions essential to the final happy result of his operations. The first bears relation to the actual state of his grapes; the second, to the state of the weather; and the third, to the state of his whole plant, crushers, presses, fermenting vats, barrels, and all other utensils requisite for a wine cellar. Any carelessness about any of these indispensable requisites may give rise to disturbances and inconveniences better to be avoided.

As to the state of ripeness in which the grapes ought to be in order to yield good wine, it is agreed on all hands that this occurs just when the grapes have attained natural—that is to say, perfect—ripeness. But no uniform indication of it can be given, and many considerations have to be taken into account to render it practically useful.

Bechamp distinguishes two kinds of ripeness in grapes—physiological ripeness, and conventional ripeness.

1. Physiological ripeness is complete when the seed of the grape is ripe enough to reproduce its kind. This ripeness may be insufficient for the production of certain wines, and even to give to all the highest degree of perfection of which they are capable.

2. Conventional ripeness, is relative to the kinds of, and the quality of, wine sought to be made from them; and in this way we can draw a line between grapes properly ripe and grapes excessively and overripe. The grape is well ripe for making good wine when the principles contained in the berry, and have to furnish the must, are in a state of equilibrium — that is,

when the sugar has attained the maximum of which the kind under consideration is susceptible, without the berry having lost its plumpness and freshness.

The grape is overripe, or excessively ripe, when it begins to show that it is passing on to become a raisin, the berries losing a part of their natural water. From this loss of normal water there results an increase of sugar in the must, and a modification of the taste of the fruit.

For the making of nearly all wines, but most especially for good table wines and food wines (*vinhos de pasto*), the first consideration is that the grapes be naturally ripe, neither over nor underripe. But how are we to know that the grape has attained to this state? Simply looking at grapes, whether white or black, is not enough; the taste for those who have competent knowledge of the kinds, and long practice, is certainly a more secure criterion, but of no use to those who need our advice. To such we can only offer the advice, touching this characteristic that they should study and compare the flavor and peculiar sweetness of each kind at the different periods of their ripening, because in nearly all of them there are appreciable differences. As for the rest, in the well ripe grape, the foot stalk and all the woody parts of the

bunch begin to shrivel, and to assume a darker
color, and the berry becomes easily detached
from the peduncle, leaving attached to it a part
of the cord which had served to nourish the
fruit. The skin of the grape becomes thinner
and more delicate; the berry on the under side
becomes nearly translucent, and appears somewhat soft; sweetness becomes the prevailing
taste, and the juice is thicker, and feels sticky
to the fingers, like sugar syrup. Analysis forms
the most reliable ground for judging when grapes
are in the condition to form the best wine; nevertheless, though there is little difficulty about
it, still but few are able to avail themselves of
its advantages. The principal ingredients whose
quantity it is most desirable to determine in the
must, are the sugar and acids, because from their
mutual relations we may deduce with much prob-
ability what the quality of the wine will be. A
high degree of saccharine implies a low degree
of acids, and musts in this condition will yield
fine spirituous wines. In this case the musts
which contain more than 25 per cent. of saccharine and less than 0.5 per cent. of acid are
represented by monhydrated sulphuric acid
$S O_3$, $H O$. When the amount of acid comes
near 1 per cent., the quantity of sugar is almost
always below 15 per cent., and the wines resulting are generally weak, cool, and but little
spirituous.

To the different kinds of grapes grown in the same locality, there correspond, when in a state of perfect maturity, different quantities of sugar, of acids, and of other principles which enter into the composition of their juice; quantities which, however, may vary within certain limits, according as the seasons were more or less favorable. But still what is of paramount importance to the vinyardist, is to know when his grapes are ripe enough for the vintage — when their sugar has attained the maximum of naturally ripe grapes, relatively to each kind, or a general average over all, when the different sorts are not separated. When the sugar appears to be stationary for a day or two, the vintage may be begun.

The plan of this short treatise does not admit of an exposition of the analytical methods mentioned above, nor in fact does the greater part of the vineyardists, to whom I dedicate this work, possess the knowledge and appliances requisite for conducting analyses. I will merely mention, incidentally, that in the first part of his "Technologia Rural," J. I. Farreira Lapa has described the instruments and processes indispensable for conducting such analyses.

There is at present in common use a special aerometer called *glucometer, gleuco-œnometer,* or simply a *must weigher*, which may, when more exact methods are not available, serve to indicate the approximate saccharine strength of musts. The way to use it is very simple. The grapes whose must we wish to test are pressed; the mass is then to be filtered through a clean linen or cotton cloth, and collected in a tall vessel, so that the instrument can float freely. If the temperature of the air and must exceed 61° F., it is necessary to place the vessel in cold water till this temperature is attained; then introduce the glucometer into the fluid, which will sink till it becomes stationary. The exact degree indicated is then read off, which (after having subtracted one degree for each twelve marked) gives the quantity of alcohol which the sugar contained in the must will yield. By making this experiment with the same kind of grapes, or with different kinds mixed in the *same* proportions, for a few successive days, it will be seen that the quantity of sugar increases as ripeness progresses; but, on arriving at a certain point, the indicated strength will appear stationary on two or more successive occasions, and this indicates that the vintage should begin.

The mere observation of the density by Gay-Lussac's densimeter, which gives the weight of a definite volume of must, furnishes a safe indication for determining the richness of a must. The denser the must the more sugar is there in it. The indications of this instrument are very important in fixing the time of the vintage and foreseeing the richness of the wine about to be made.

In a vineyard planted with one kind only, and in which all the conditions of soil, situation and exposure should be uniform, one would say that the grapes would all arrive at maturity at one and the same time, but this very seldom happens. In most of our vineyards, besides the diversity of natural conditions, the kinds culti- vated are very various, and as each one has its own period of perfect maturity, there is no hope of all of them being ready for gathering at the same time. It is necessary, however, in the greater number of seasons to make the vintage all at once, and to this end we have to select a time when the greatest part of the grapes are quite ripe, hoping that if some part of them be overripe they will be compensated by others not fully ripe, to the extent of their deficiency of sugar.

In this as in most cases, a middle course is the best. If we gather the grapes before sub- stantially the whole of them are naturally ripe, we incur the risk of having a weak, acid, or a green, rough wine. If, on the other hand, we

wait till the whole have gone beyond natural maturity, we may reckon upon having a thick must
difficult to ferment, and one which, with much
care even, can hardly ever be properly made,
being all the time liable to a succession of
changes which may ruin it. Only certain special
kinds of white wine require the grapes to be excessively ripe, but the processes of making them
are quite different from those employed in the
making of genuine, alimentary (*alimenticios*)
wines, such as ought chiefly to occupy our attention. In the case of making dry wine, it is
far better to gather the grapes a little before
full natural ripeness than allow them to pass
beyond it. The reader will now see that I am
not treating of the making of those generous
wines which are made after the pattern of those
of the Douro, in which, through the addition of
alcohol, the changes liable to arise out of the
excess of sugar in the presence of matters liable to cause fermentation, are effectually prevented.

Meteorological conditions of the atmosphere
exercise much influence upon wine-making, and
deserve attention in this place while I am treating of the vintage. Clear, dry, and moderately
dry weather is best for the vintage. Warmth
favors and stimulates fermentation; cold op-
poses or retards it; consequently the tempera-

ture in which the grapes arrive in the cellar has
a powerful influence on the rise and progress of
fermentation. Damp, moist air has a not less
remarkable influence. In vineyards where the
kinds of grapes do not form much sugar, or
when they are prevented from attaining to perfect ripeness, dew, damp fogs and rain, through
damping or wetting the grapes, tend strongly to
dilute the must and to leave the wine watery
and flat. On this account all gathering should
cease as long as the dew or fog or rain is upon
them. In vineyards, however, where maturity
is excessive rather than deficient, the moisture
of a fog does no harm. In gathering grapes for
white wine, it is even advantageous to gather in
foggy weather, since experience has proved that
in this state they yield wines which clear themselves more easily and more perfectly. Rainy
weather is always prejudicial during the vintage.

The judicious and experienced vineyardist,
bearing in mind all these circumstances, can
lay out the plan of his vintage, and will almost
always carry it on and finish it in good condition. In view of the state of maturity of his
grapes, and of the meteorological circumstances
of the atmosphere, he can direct his work and
foresee the results of his crop. When grapes,
as often happens, ripen very irregularly, some
being quite ripe and others far from it, it be-

comes desirable to make partial gatherings, selecting only the ripest first, to make a wine of
the first quality, and taking the rest as they
ripen and making a second-class wine. These
partial vintages often become very advantageous, especially to those who have knowledge
and experience in making fine wines.

It may, and it nearly always does, happen,
owing to accidental circumstances, that among
the grapes of the same vineyard, we meet with
more or less which are unfit for making good
wine, either through being green and imper-
fectly formed, or dry and shriveled, or rotten
and moldy, or injured by hail, without men-
tioning those affected by oidium. Now, it is
simply indispensable that every grape found affected in any of the above ways should be picked out. Not one of them must ever be allowed
to enter the crushing room or the fermenting
vat, where the object is to make good wine,
under pain of introducing into it causes of certain destruction, which will show themselves
sooner or later. Green and dry grapes impart
to wine roughness and a detestable taste. The
rotten and mildewed, and all such as have un-
dergone any change, over and above the bad
taste they impart to the wine, bring along with
them into it the germs of ulterior fermentation,
which eventually ruin it. The only good to be

got out of them is by pressing them apart and fermenting their must separately and without the skins, and so obtaining an inferior wine. Thus the mildew, etc., will be arrested in the pomace.

From the period when oidium attacked our vineyards, and when we had to employ sulphur to arrest its ravages, it has happened not unfrequently that a notable portion of the sulphur found its way into the press-room, which, during fermentation, gave rise to sulphydric gas, and which remained in the wine, communicating to it an offensive taste and the smell of rotten eggs. In some cases this trouble may be avoided when the grapes are well formed and very saccharine, by subjecting them to careful washing. I advice, when such a washing is needed, to proceed as follows: In a huge tub full of water, which is to be frequently renewed, I cause the wicker baskets in which the grapes are brought from the vineyard, to be immersed and carefully shaken about for a few minutes. The sulphur adhering to the bunches soon washes off and falls to the bottom of the tub through the interstices of the baskets. Afterwards they are allowed to drain for awhile and lie exposed to the air, and are then ready for the press-room. Wine so made never gives any indication of sulphydric acid gas.

2

One of the requisites of a good vintage is, that it should be carried on in such a manner as that each fermenting vat should be full at the close of a day's work, so that fermentation, etc., might go on with uninterrupted regularity. Vineyardists who neglect this condition commit a grave error, giving cause for an irregular fermentation, which always injures the quality of the wine. As a matter of course, the number of hands employed ought to be calculated for this end, and the work directed from this point of view. In the gathering and selection of the grapes, all possible care and vigilance are necessary. The employment of proper scissors is a thousand times preferable to the vile knives in common use among our vignerons. With the scissors the bunch is cut off without injury and without shaking the vine, which is not the case when ordinary knives are used.

THE PLANT REQUIRED FOR MAKING WINE.

Before treating of the different operations of wine making, it may be useful to pass in review the utensils and appliances required in the work.

In the wine industry, as in most others, the appliances used have much to do with the after products. In wine making, the *process adopted* is as important as are the vessels, the apparatus, instruments and utensils which must of necessity be employed.

The simplicity of the plant is a very praiseworthy matter, so long as it is sufficient for the prompt and effective execution of the labor.

In the greater part of the chambers in which our wines are made, which are our wine-workshops, the simplicity is excessive and almost primitive, so much so that the same word serves to designate the house or office, and the stone tank in which the grapes are trodden, and which forms their principal part.

A quadrangular stone tank, roughly built, raised somewhat above the floor of a house, nearly always without a roof; another tank of much smaller dimensions contiguous to and somewhat below it, which we call *lagarica* or *pio*, a beam fixed at one end against the wall of the tank, and having at the other end a hole into which is secured a smaller beam loaded with one or more huge stones, the whole forming the wine-press, constitute the wine-making appliances of the greater portion of our wineries in the provinces of the north. The keeping-cellar is usually in no better condition.

In the great vineyards of the Douro, and in others formed after their pattern, the system is the same, with the essential difference of the size, number and better disposition of the parts. There the press-houses and cellars are vast, regularly built and contiguous to each other; the first being erected on a floor higher than the second, so as to admit the wine to run from the press-rooms to the great vessels through covered channels.

The tanks, the wells (pios) and the presses are placed methodically and symmetrically in the press rooms, as are the great keeping-vessels (*toneis*) in the cellar.

To each tank there is a corresponding window through which the grapes are poured in as they are brought from the vineyard. To each two tanks there is a well which receives alternately the wine made in them, and from which it runs through the covered channels into the tuns (*toneis.*)

All these arrangements would be highly worthy of imitation if they could without inconvenience be adopted in the making of all wines. The wines of high and rapid fermentation, made with grapes very rich in sugar, and at the expense of a prolonged and violent treading by men, and whose density (*expessura*) is corrected afterwards by the addition of brandy, can be made only in (*lagares*) stone tanks, such as those of the Douro.

On the other hand, the more natural, pure, genuine wines, which do not require to be either excessively trodden or derated, and above all, ordinary wines from grapes of moderate or very limited saccharine strength, cannot be conveniently made in these shallow stone of little depth and broad surface.

There are many reasons why these tanks (*lagares*) should be given up; they are difficult to wash perfectly clean, on account of the roughness of the stones, whilst absolute cleanliness of all vessels in which the wine is made, is an indispensable condition of the healthy after formation of the product, because they cool the must and waste the heat required for fermentation; because their great opening and shallowness prevent the must and pomace from being protected from the air, which might prove very injurious, as we shall soon see, to wines of less active fermentation, and therefore proportionately more prolonged; because on account of the very nature and form of their construction they occupy much space, because the angular spaces or corners interfere with the uniformity of the fermentation, withdrawing a portion of the must from the general movement to the prejudice of the regular formation of the wine, and lastly, because, when out of any motive we may wish to alter their dimensions or change the system, we are put to much more expense than would be the case if we had to alter wooden vessels and appliances, or dispose of them, as they always possess an appreciable value.

In those countries where the best food wines (*vinhos de pasto*) are made, stone tanks for working the grapes are wholly unknown; and none are used but such as are made of oak hooped with iron; and these realized all the requisites for good vinification. This kind is not wholly unknown in Portugal, but is by no means common; and those that we use are not always of the most convenient construction.

Wine making in stone tanks, were it not for the drawbacks already pointed out, would realize the inestimable advantage of extreme simplicity, for the entire work could be done in them; the stripping, if desired, treading, fermenting and pressing the pomace. The great vats (*balseiros*) on the other hand, serve only for fermenting. The grapes, when put in must be completely crushed, and as soon as fermentation is over, the pomace must be taken out to be pressed. But notwithstanding this apparent round-about way, they are in every sense to be preferred in the making of pure, genuine alimentary wines, principally the red kinds; for white wines, not intended to be blended with red, may be made differently, as we shall see hereafter.

[N. B. Here occur three paragraphs touching the construction of cellars and presses, but they contain no hints which are not already generally reduced to practice in California.]

It is of the very utmost consequence that the crushing and pressing room should be well ventilated, and kept perfectly clean; the floor laid with brick or cement, the walls whitewashed with lime, the roof lined, and the windows glazed. Every sort of vessel, instrument or utensil must be, at any rate during the vintage, faultlessly clean. It is hardly necessary to say that the same applies to the cellar and every vessel in it.

PREPARATION OF THE GRAPES.

When the grapes have been delivered at the crushing room, selected and freed from all defected ones, they should be reduced to must as soon as possible, but under varying circumstances, and in different localities, they are submitted to a previous preparation, which, though not always an advantage, is sometimes useful. This is known as *stripping*, the separation of the berries from all woody matters, which is commonly called pomace.

In some countries renowned for their fine wines, stripping is had recourse to as a general rule; in others not less famous, it is absolutely condemned, and employed only on very exceptional occasions; in others a middle course is adopted, the stripping being either partial, total or not at all, according to circumstances. Results obtained in different countries seem to justify these different methods; so there is nothing absolute in the theory of stripping. The discussion of the theory of stripping does not enter into my plan. I aim only at pointing out certain indications which may prove practically useful.

Stripping may prove either useful or injurious to the fermentation, either chemically or mechanically; chemically on account of the nature of the matters supplied to the must; mechanically by reason of the division and volume which it makes in the fermenting vessel.

In France, Germany, and Hungary, white wines are fermented, as a general rule, after complete separation of the berries from the stalks. In making such, the presence of pomace during the fermentation is neither necessary nor useful, but on the contrary prejudicial; but in this instance there is no need of stripping, since the grapes are put through the press just as they come from the vineyard. The presence or absence of the stalks is a question affecting red wines only.

The stalks, on account of the substances which they contain, may be liable to supply a considerable amount of support to the fermentation; and from this point of view their utility cannot be doubted in the case when grapes of high saccharine strength are deficient in nitrogenous matters, acids and salts, as often occurs in our warmest and best situations.

On the other hand it might impart a certain amount of acerbity and astringency to wines made from thin, watery musts, which would improve their bouquet and aroma. Still, this

2A

greenness and astringency may easily become excessive, and leave the wine rough and harsh, unless care be taken not to have it very long in contact with the must, or if the fermentation be unusually prolonged, or when we desire to obtain all the body and color possible.

The stalks may also be useful in the fermentation of musts very thick and rich in sugar; because they leave the fermenting mass more porous and more accessible to air, and so promote its progress.

Experience has proved that in most cases, wines made with the stalks are less liable to go wrong and are of greater durability; and that the roughness which they show at first, gradually passes off, and eventually disappears altogether.

Some writers on wines attribute this power of endurance to the presence of tannin in the stalks; others deny the existence of tannin in the woody matter of the bunch of grapes; and convinced of its absence, consider the presence of the stalks useless. The chemical nature of the stalks has never yet been properly studied; still the experiments which I have made upon it, satisfy me that tannin, properly so called, does not exist in the stalks of the bunch, like it does in the husks and seeds of the grapes. In the meantime, though the astringent principle of this part of the bunch does not strike the per-salts of iron black, as does that of the seed, it by no means follows that it may not exercise a preservative power in the wine, analagous to that which true tannin does, undoubtedly.

From all the above, we may conclude that there is no need of laying it down as a rule, that stripping should always be adopted in making pure, genuine red wine; but rather that we should pay attention to the condition and nature of the grapes, and to the quality of the wine we desire to make, and guide our operations accordingly.

If the grapes of their own nature, and on account of the conditions of the season, are very ripe and sugary, so as to yield a very dense must, stripping will be out of place.

If they are watery, deficient of sweetness, flat-tasted and insipid, even in this case the stalks will help to give bouquet and aroma.

If they taste harsh and notably acid, it is desirable to make a partial stripping, and occasionally a total.

If owing to exceptional circumstances, it should be evident before the vintage, that a large portion of the berries would have to be rejected, in this instance it becomes desirable to make a partial stripping.

Of course in practice many circumstances

may occur to necessitate a modification of these precepts. It may, for example, happen that the vintage may be so very abundant as to exceed the capacity of our fermenting vats, in which case stripping would be useful by reducing considerably the volume of material to be dealt with. The construction of the fermenting vats calls for attention also. If they be covered the stalks do no harm in most instances; on the other hand, when the fermentation is made in uncovered ones, the stalks are useful in forming " the cap," which rises to the surface of the must.

When our aim is to obtain a soft, velvety wine, or one somewhat sweet, stripping wholly or partially is desirable; if, on the contrary, we desire a rough, harsh wine that will lay hold of the tongue and throat, then not only no stripping, but prolonged fermentation on the skins and stalks. Rapid fermentation, and brief contact of the stalks with the must impart to it no excessive roughness or astringency.

In view of what has been said, stripping should be had recourse to or not, wholly or partially, according as we note the various conditions of the state and nature of the grapes, and the quality of the wine we desire to make.

Stripping may be done in many ways. The most simple is to use a wooden rake upon an

inclined plane over the crusher, which will easily detach the berries. This plan has the advantage of taking off only the ripe berries, leaving the green ones adhering to the stalks.
There are many forms of sieves which can be
made to answer the purpose well. As the
bunches are worked backwards or forwards over
the wires, the berries drop through the interstices and are then crushed. On the Douro
they employ huge frames with wire bottoms
which they call *escangalhadeiras* (breakers up),
because they separate the refuse or stalks. In
Burgundy, where stripping is rarely practiced,
even partially, they employ a sort of wicker
sieve placed over a small barrel, into which the
berries fall. It matters little what method is
adopted; the most simple, convenient and the
cheapest, is the best.

Another preparatory operation, and indispensable in making red wine, of whatever nature it
may be, is the crushing of the grapes so as to
get out of the berries whatever matters they
contain, wh·ther liquid or solid. The method
followed by *us*, of having the grapes trodden
by men with bare feet, offers great advantage
in nearly every instance, when the work is done
with necessary cleanliness, which is indispensable to keep the wine from deteriorating in its
after stages. The pressure exerted by men's

feet is enough to crush the berries, without
smashing the seeds or the stalks; while the
commotion produced in the mass by the alter-
nate action of the feet, during the operation,
forces into it plenty of air to start and continue
the fermentation.

In most cases when we start in to make genuine natural wines, it is quite sufficient that the
berries be thoroughly crushed, in order that the
must may be set free, and that contact may be
maintained during fermentation, between the
liquid and solid portions. When, however, we
seek to make the generous, strong wine of the
Douro, long and violent treading, is both useful
and necessary, to enrich the must, to ærate it,
and to ensure a more active fermentation; but
this process does not hold good for making genuine wines for regular consumption.

In making white or liqueur wines, there is no
need of treading the fruit; it is enough to crush
them with the press and collect the must in the
fermenting vessels. In this way alone can
bright pale white wines be obtained, like those
of Germany and France. In any other way,
how limited soever be the contact of the must
with the stalks and skins, the wines acquire a
yellow color, more or less intense, due in part to
the coloring matter of the skins, and partly to
the oxidation of certain matters yielded by the
stalks.

[N. B. Machinery now used as a substitute for treading seem to have failed hitherto to effect the requisite aeration of the must, without recourse to more or less efficient means, none of which seem nearly equal to working with the feet. J. I. B.]

To resume: stripping or no stripping. what is indispensable, is to effect by treading, crushing, or by any other method, a perfectly fluid and homogeneous must, and to promote contact among all its parts, whether fluid or solid, so that fermentation may go on uniformly, and the reactions, necessary to a perfect wine, may be established.

In this, as well as in all other operations in wine-making, the utmost neatness and cleanliness must be observed, both in casks and utensils; and must be most rigorously enforced on the men employed in treading the fruit. The slightest neglect in these particulars, may be the means of having germs of disease introduced into the wine, which will eventually develop and greatly injure or destroy it.

FERMENTATION.

The operations which I have described hith-
to are only preparatory to the wine making, or
the transforming of must into wine. Wine, according to the common acceptation of the word,
is the liquid resulting from what we call vinous, spirituous or alcoholic fermentation of the
juice of grapes.

Seeing that vinous fermentation is an indispensable condition in making wine, it is important that we should have clear ideas of the circumstances which favor or prejudice the formation of it, in order to properly appreciate the
value and importance of the practical rules
which are to guide us in making wine.

It is not my intention to go extensively into
the doctrine or theory of fermentation, as ac-
cepted by modern science. I shall limit myself
to pointing out only such fundamental facts as
are necessary to understand the rules and practical processes.

Alcoholic fermentation is a transforming
movement, produced in liquid containing sugar,
either natural or artificial, in presence of one
particular ferment.

The ferment itself is an organized and living *being*, which for its nourishment and development consumes a part of the sugar, transforming the rest into carbonic acid, which passes into the air in the state of gas, and into alcohol and other principles, which remain in solution and constitute the spirituous liquid.

Alcoholic fermentation is not the only phenomenon of the kind with which we are acquainted. There are indeed several different species of fermentations, all of which appear, like this, to be corresponding actions of a vital force. In all of them, one or more living beings, vegetable or animal, excite in the liquids, or mediums in which they exist, through the exercise of their vital powers, chemical reactions and modifications, in which are formed, at the cost of the existing products, other products of a quite different composition. Setting aside, however, these diverse fermentations, we will direct attention for a while to that one which most concerns us in wine making; because it is essential and indispensable thereunto. We call it the alcoholic fermentation, in which, as just said, the sugar is transformed by the ferment into alcohol, and a few other products.

The conditions essential to produce those effects are the following:

1st. A watery liquid containing a certain quantity of sugar in solution.

2nd. The presence of air.

3rd. Ferment already formed, or nitrogenous matter of the nature of what chemists call *albuminoid*, on account of the analogy of its composition to that of the white of an egg, and which is capable of affording nourishment to the ferment.

4th. A moderate temperature, 15 to 20 centigrade.

If we make a not very strong solution of sugar in water, and add a little ferment to it, whether beer-yeast, or wine lees, or in their stead the albuminoid matter mentioned above, it will after a while enter into fermentation, which will show itself by becoming turbid, by effervescence caused by the evolution of carbonic acid gas, and by a perceptible rise of temperature. During this fermentation, the sugar disappears, and we shall then find in the liquid alcohol, or spirit of wine, a little fluid substance, called *glycerine*, and a crystallizable acid—the *succinic*. If the nitrogenous matter or ferment had been supplied in sufficient quantity, there would be found, when fermentation had ceased, a precipitate containing the organic being formed during fermentation. When it becomes still, the liquid gradually clears itself.

Now, if the quantity of ferment or nitrogenous matters, at the expense of which the transformation is carried on, were sufficient, the whole of the sugar would have disappeared; if, on the other hand, the ferment was not sufficient, and the liquid deficient in nitrogenous matters, a portion of the sugar would remain undecomposed. If the nitrogenous matters were in excess, the surplus would be found in the lees.

What happens in an artificial solution of sugar, gives us an idea of what takes place in liquids containing natural sugar, such as the must of ripe grapes, when placed under similar conditions. The must of ripe grapes contains water, sugar, and both acid and neutral vegetable matters, and mineral and vegetable salts. Roughly estimated, we may say the water of the must will amount to 80 per 100; the sugar may vary from 15 to 30 per cent., and rarely more; and the rest of the matters rise not unfrequently to as much as 30 per cent. of the total weight of the must.

Among the neutral matters, there is always present a portion of nitrogenized matter necessary for the life and propagation of the inferior organism which sets the alcoholic fermentation in motion; and among the salts, bi-tartarate of potassa, which is met with afterwards as a deposit from the wine.

In order to cause fermentation in a solution of sugar in water, it is necessary to add ferment, or yeast, to it, or nitrogenous, and to place it in contact with air, at the temperature mentioned above; but in order that the must of grapes should enter into fermentation, all that is required is to leave it at the temperature referred to in contact with air for some time. It looks as if in this case the fermentation took place spontaneously. In the theory generally received at the present time, of the vinous or alcoholic fermentation of the juice of ripe grapes, it is admitted that the germs of the organism, which develop in the liquid, as long as fermentation goes on, are carried in the air, like other germs, for instance of mould, of oidium, and a thousand other vegetable and animal organisms, which we see developing every day in circumstances favorable to their growth.

These germs, finding the sugar and nitrogenized or albuminoid matters ready dissolved in the water of the must, and under favorable conditions of temperature, develop, as if in a fertile field, and thrive and propagate in the liquid at the cost of the sugar and nitrogenous matter. The sugar, by yielding part of its elements to the ferment, is transformed into carbonic acid gas, which escapes in the well known bubbles, into alcohol, into glycerine and succinic acid, which remain dissolved in the liquid along with other principles, neutral, acid, saline and aromatic, which constitute the wine. The nitrogenous matter serves as plastic nourishment to the organism, which at the end of the fermentation, and when all the sugar has been broken up, settles down with the solid matters, which were held in suspension in the must, or rises as scum to the surface.

Rigorous experiments, and which any scientific chemist can readily verify, convinced Pasteur that 100 parts of pure grape sugar yield by fermentation the following:

Carbonic Acid	46.67
Alcohol	48.46
Glycerine	3.25
Succinic Acid	0.61
Matter yielded to the ferment	1.03
	100.00

The sugar of the grape is the essential part of the must which goes to form the fermentation of the wine; and as the amount of it is greater or less, so will be the alcoholic strength of the wine. Nevertheless all other substances contained in the tissues of the grapes, in the skins, in the pulp, in the seeds, and even in the stalks, contribute to modify the quality of the liquid resulting after fermentation. Thus we learn that wine is not a simple, but a very complex liquor, the qualities of which depend on the nature and richness of the grapes of which it is made, and on the circumstances under which it is made; for these may be such as to insure its keeping and improvement, or accelerate its ruin.

The above brief exposition of the theoretical principles, seems to me sufficient for the understanding of what occurs during the transformation of must into wine; and to awaken the attention of the vintner to all the circumstances and conditions surrounding the operation.

We note two different methods of wine making. It is done either by separating the must from both stalks, and skins, and seeds, and this is the simple way in which French and German wines are made; or the fermentation is conducted in the must, together with the whole or a portion of the solid parts of the grapes, as is universally the case in making red wine.

In the first instance vinification is very simple. The grapes, picked and clean, are crushed right away and the must collected; the pressing made and the juice added to the must, and put at once into the vessels in which it has to ferment; the bung-hole being left open for the escape of the carbonic acid gas, and scum which brings over the part of the ferment which rises.

The must contains in itself all that is requisite for making those wines. During the crushing and pressing, the air deposits in it the germs of alcoholic fermentation, and by being brought into intimate contact with it, prepares it for the subsequent reactions about to take place. Thenceforth, under suitable temperature, fermentation is established, and continues of itself, and gives no further trouble, since the carbonic acid gas, which is evolved, protects it from contact with the air by forming as it were, a gaseous atmosphere over it.

In the second case, in which, for the production of red wines, it is indispensable that the must be fermented in contact with the skins, chiefly that the alcohol, as it forms, may dissolve out the coloring matters which have been formed on their inner surfaces, the operation becomes complicated, and to be successfully conducted demands many and serious precautions.

WINE MAKING IN PORTUGAL.

First of all, let us take a view of the method, in common use in Portugal, of making wine in stone tanks (*lagares de pedra*).

Generally the grapes are not stripped, but are put into the tank (*lagar*), spread out and bedded, till it is a little more than two-thirds full, in order to leave sufficient space for the cap when it rises, so that it may not run over. As soon as all the grapes are in the *lagar*, the men with bare feet and legs, well washed, go in and begin to crush and tread the berries, treading first with one foot and then the other. This work continues for a sufficient number of hours, and, if the temperature be favorable, fermentation sets in, and may be observed in the appearance of bubbles of gas that form on the surface, by the increase of temperature and by the wine-smell, which begins to be perceptible. At the same time the empty skins and the stalks keep rising to the surface, lifted by the gas caused by the fermentation, and, if treading be interrupted for a time, the surface of the must will be covered by a thick cap of stalks and skins. Fermentation goes on more rapidly in the upper part near the cap, than in the lower, where there is simply must. The more active of chemical reactions, and of vital functions, which are correlative to chemical action, are always, and under all circumstances, accompanied by elevation of temperature. Through the labor of treading, the solid and liquid portions of the grapes become intimately mixed, and by the continuous movement of the whole mass in contact with air, the air penetrates every particle of it, depositing in it the germs of ferment floating in it, at the same time supplying the oxygen which appears to be indispensable to the development of the ferment, as well as to the oxidation of the coloring matters, and probably other reactions but little known. When treading is finished, and the cap has formed over the surface of the liquid, the air acts only on the porous surface of the cap, more particularly if the tank be very full, and an atmosphere of carbonic acid gas cannot form, and remain tranquil over it, to protect it. This contact of air with the cap, if continued for a considerable length of time, may cause, as we shall soon see, very ruinous reactions.

When the fermentation has been regular, and made under favorable conditions, after some hours of tumultuous action, it slackens; the "cap" begins to dry and cool; to crack and sep-

3

aràte from the covers, and to show signs of
sinking. At this point the wine is generally
considered made; its smell has become vinous,
its taste somewhat astringent, the sugar has
disappeared, its density much below that of the
must, and nearly that of water. The time has
now arrived for drawing off the wine, completely free from any part of the cap, into clean vats
ready to receive it. As soon as the tank is
empty the cap is pressed, and the pressings
added to the rest of the wine.

The above is the regular order of operating.
We will now examine the drawbacks and advantages which it presents in the case of ordinary wines of consumption, or more properly,
genuine wines for daily use.

Considered apart from surrounding circumstances, the whole operation is quite simple.
Into one and the same tank, the grapes are
emptied as they arrive from the vineyard; here
they are trodden; here they are worked and
aerated; here they are fermented; and here
the pressing of the pomace is done. If this
work has been well done, with all cleanliness; if the weather was all that could be desired; if the fermentation went on quickly and
regularly without interruptions, and finished in
a short time, not taking more than two or three
days, we may conclude that the wine made will

prove as good as we have any right to expect, regard being had to the grapes and the year.

But supposing the fruit to have been gathered in cold damp weather, the chilliness of the stones of the tank will be added to the chill of the air; and still more, the large open surface of the tank will retard alcoholic fermentation; and then there may be seen in spots that glutinous, sticky product of bad fermentation, so injurious to the wine.

Supposing the alcoholic fermentation to have been established; if, for the above or other causes, it be much prolonged, it slackens, stops, and becomes interrupted, contact with the air sets up other ferments in the "cap," chiefly the *acetic*, which converts the alcohol already formed, into vinegar; or may be, the *putrid* fermentation, in which vibriones and other inferior animalcules are generated. In whichever case, whether the cap be pressed down through the liquid to increase fermentation, or to obtain more color and body, or whether we simply add the pressings of the pomace to the rest of the wine, we cannot help introducing destructive germs, which sooner or later will show their effects to the great injury of quality and keeping power.

Again, supposing the fermentation good and regular; since the surface of the cap is large, which is exposed to the air, and the temperature high, not unfrequently as high as 85 to 90° Fahrenheit, a portion of the alcohol and aromatic principles is lost by evaporation in prejudice of the richness of the wine. Injuries of this kind are always to be regretted, but more especially in the instance of weak wines. Of course, when the wine intended for brandy, any loss of alcohol is a *dead* loss.

The tastes of consumers and the exigencies of commerce, require very frequently full bodied, highly colored wines. To satisfy these requirements, it is necessary to keep the must mixed with the pomace a long time. Now, for reasons already assigned, tanks (*lagares*) are wholly unsuitable.

Grapes extremely ripe, very rich in sugar, and yielding very thick, dense must, such as are met with in the best localities of the Douro, make good wine in those stone tanks, for they require excessive treading and agitating in order to render the mixture of solid and fluid portions perfect; also that the air may penetrate it, and facilitate the reactions, and to cause the fermentation to proceed with ease and energy. But wine so made is too rich and too full-bodied, and only becomes perfect after a length of time, on which account it is necessary to dose it with brandy, to protect it against injurious fermentations. This dose of brandy is put into it when it is vatted, or in the month of March, before the temperature of the air is sufficiently warm to set up fresh fermentation. No such process can be used in making natural wines for use as food, which should not be · highly charged with flavoring matters, but light and dry, and which should be fit for use in shorter time.

METHODS OF THE MEDOC AND BUR-
GUNDY RECOMMENDED.

A method of wine-making, approaching as nearly as possible to that in use on the Medoc, by which the excellent wines of Bordeaux are made; or the Burgundy method, which also produces wines of great merit, seems to me to be perfectly well suited to the larger portion of our viticultural centres, where table wines of a superior order are possible to be made.

In both the above named viticultural centres of France, great wooden vats, holding from 900 to 4000 gallons, in the shape of truncated cones, are in use. These vessels have their bottoms considerably broader than the tops. The grapes, whether stripped or not, in whole or in part only, are first crushed and trodden, either by crushing machines, or by men's feet. The must and skins, etc., are then put into these vats till they are conveniently full, leaving 12 or 15 inches, to avoid any overflow during fermentation. This filling of the vats should be done with the least possible delay, and never allowed to exceed one day, in order to prevent the fermentation, when it has once begun, from being interrupted, for it always injures the wine.

As fermentation commences the pomace rises, the same as in the case of the tanks, and forms a layer on the top which they call the "cap." Carbonic acid gas is thrown out in torrents; and the motion of the liquid, due to the reactions taking place in it, becomes at first tumultuous, then after a while subsides, and after the lapse of a longer or shorter space of time, ceases entirely; and the pomace which formed the cap, cools, contracts, and begins to fall. The wine is now made.

During the period of tumultuous fermentation, the wine makers of Burgundy, especially those who employ open vats, submerge the pomace in the must once or twice a day, to equalize the fermentation, but more particularly in order to obtain a maximum of coloring matter in the wine.

The above is the simple and regular order of operating, but it is always liable to the same dangers, as I noted in the case of tanks, so far as concerns contact with the air.

If all the circumstances are favorable, the grapes perfectly ripe, the weather uniformly warm, dry and clear, fermentation sets in immediately, runs its course rapidly, and at the end of three or four days, is finished and the wine made. Under such conditions, the mixing of the cap through the must, in order to equalize the fermentation and obtain more coloring matter for the wine, can in no way be injurious to the wine. But when the state of the weather is less favorable, when fermentation is retarded, and evil fermentation shows itself on the cap, as the acetic, or putrescent fermentation, then become apparent the drawbacks to this method.

To obviate these dangers, some wine makers have, since a long time, adopted plans of covering the vats to protect them from contact with the air during fermentation. Many plans have been recommended and put in practice, to conduct fermentation in closed vats; and in this, as in other cases, there appears a good deal of exaggeration. Absolute covering, hermetically sealing the vats, is impracticable and absurd, for there must be an escape for the carbonic acid gas, or else it would burst the vat. A perfect covering, with a safety valve, and with a tube for the escape of gas, dipping into a vessel of water, if on one hand it overcomes the contact of air, without risk of bursting the vat, it on the other hand exerts a pressure on the fermenting fluid, and checks its regular progress, not without risk of injuring the quality of the wine, and also at the same time increasing the

cost of production. An ordinary wooden covering provided with holes for the escape of gas,
and just laid over the fermenting vat, or even
loose boards, covered with matting in a similar
way, as soon as fermentation has fairly set in,
presents no inconvenience, and completely sat-
isfies all requirements.

On a supposition that the rapid dispersion of
carbonic acid gas, assisted by the elevated temperature of the fermenting must, would cause a
pure loss of alcohol and aromatic elements, different plans have been suggested for condens-
ing and retaining these elements. Such were
the old apparatus of Gervais and the modern
one of Minard. The first one proved to be
practically useless, and was abandoned. As to
the other, but few comparative experiments
have been made, and while some consider it
highly advantageous, others of not less authority declare that its advantages are not an equiv-
alent for the trouble it causes. For myself, I
neither advise nor condemn its use.

In view of what has been explained, I consider the best course for our people to adopt in
making natural, genuine, nutritious, table
wines, is to ferment in moderate sized vats
(*balseiros*), simply covered.

To recapitulate—the grapes clean, and free
from imperfect berries, stripped from the stalks

3 A

wholly, partially or not at all, conformably to
the doctrine above laid down, should be thoroughly trodden in the tank, or on a suitable
platform or bench, placed at the side of or over
the vat, so that the crushed mass may run in,
until it is within 12 or 15 inches of being full.
This work ought to be done in the shortest
time possible, so that no vat should be left to
the following day unfilled and unfinished. The
next thing to do is to *stir up* and *thoroughly mix*
the whole mass, with rakes of wood, or other
suitable wooden implements, to render it uniform and give it all the airing possible; and as
soon as general fermentation has become established, to cover the vat and leave the fermentation to run its natural course. As soon as fermentation has ceased, and the wine appears to
be what was desired—that is, when the sweetness has disappeared, and its place has been
taken by the taste of wine, and moderate astringency and peculiar roughness of tannin, which
is necessary for its keeping, and when its color
is satisfactory, it is time to draw it off into
keeping vats. A densimeter or saccharometer
will also afford information if the wine is made;
for if it is, the instrument should stand nearly
at the point of *water*. When the wines are intended to be quite dry and hard, the instrument
should indicate zero, 0. Such as are meant

somewhat soft, may be drawn off when the instrument marks 2, 3 or 4 degrees above zero, so as to admit of slow fermentation in the cask.

Before drawing the wine out of the fermenting vats, it is a custom in wine countries, which employ the method just described, to stir the cap through the wine and mix it well, so as not to lose what has been soaking the skins and matters forming the cap, always richer in body and alcohol, with the rest of the liquid in which the fermentation had been slower. There is nothing wrong about this practice; it is rather advantageous, when the condition of the cap is satisfactory; that is, when no injurious fermentation has taken place in it; and when the smell and taste of it are plainly those of wine.

On account of the fact that the most perfect and profitable way of making red wines is to ferment on the skins, many writers upon wine making have recommended special plans for keeping them submerged in the must during fermentation. This is effected by a grate, or a second covering bored full of holes, which can be fixed in the fermenting vat at some convenient depth. This, of course, will arrest the skins, etc., as they rise, and allow free passage for the liquid, and for the escape of gas. Some simple mechanical contrivance to secure it in its place against being lifted up by the pomace

will be needed, but can be easily provided.
The permanent depression of the pomace below
the surface, yields excellent results, and saves
labor, as to the quality of the wine. This idea
has been amplified by a proposition to use sev·
eral such perforated lids or grates so as to dis·
tribute the pomace as equally as possible
through the fermenting must. Though not yet
in common use, this idea is being carried into
practice at the present time.

In places where the grapes ripen well, and
where the climate is favorable for a prompt and
complete fermentation in a short space of time,
open fermenting vats, or only slightly covered,
with the pomace kept below the surface, may
be safely used. But when we cannot reckon
upon the completion of fermentation within
the short interval of three or four days, it is
safer and better to use the covered vats describ
ed above, those that protect the cap from the
air, and allow the escape of the formed gas
and check the cooling of the fermenting mass.

In the making of fine, soft, delicate red wines
for the table, the vinification in (*balseiros*) moderate sized vats is the safest; but when we treat
of ordinary wines of consumption, the process
admits of much modification, for then we may
use the huge vessels with wide top openings to
receive the must and pomace. In this instance

the grapes are pressed in the (*lagar*) tank, or on a suitable bench or platform, and the must and pomace run into the vessel till it is more than two-thirds full, leaving plenty of space for the cap not to reach the outer edge of the opening. Under these conditions, the fermentation will be slower; but since the accumulation of carbonic acid gas in the upper part of the vessel impedes the contact with atmospheric air, those injurious fermentations cannot occur, and the fermentation may go on for a longer time without any other drawback besides the increase of the body and hardness of the wine, matters which might be reduced by a previous partial stripping of the grapes.

It very often happens that the conditions of the weather are such as to so interfere with the alcoholic fermentation as to prevent its ready development. In most instances it will have to be assisted with artificial warmth. The readiest way is the cautious warming of a portion of the must, taking particular care lest it should get a bad taste. This heated must, never beyond 175° Fahrenheit, is intimately mixed with that in the vat. It is possible to devise other methods of warming the musts, but all appear more or less risky. When there is at hand a heap of pomace quite sweet and fresh from a recent pressing, and not as yet alter-

ed by the action of the air, we may with advantage mix it with the must, and being warm, it
will directly set up fermentation. Undoubtedly this is one of the best means of pro-
moting fermentation. But suppose we have
no pomace, in the condition just stated,
we can supply its place in the following
manner: place a small cask in a place in the
house where the temperature is over 62° F., and
start fermentation in a quantity of selected and
well crushed grapes, which should be in some
relation to the size of the vat which we have to
warm; for example, for a vat of 1100 gallons, it
will suffice to warm the must to the above mentioned heat of from 50 to 100 gallons of must.
Then when the fermentation has become very
active in this warmed must it should be added to
the large body, and the fermentation will go on
with regularity. Since alcoholic fermentation
needs a moderate temperature, higher than 62°
F., and knowing that any circumstance which
checks and reduces it must be injurious, it becomes a matter of necessity that the fermenting
cellars should be kept in good repair, and protected against changes easily caused by cur-
rents of cold air. It is, however, only proper
to observe that this arrangement is inseparable
from one great inconvenience, viz: injury to the
health of the workmen, who have to labor in
cellars where vinous fermentation is going on.

Carbonic acid gas, evolved so abundantly during fermentation, is a deadly, irrespirable gas, which is followed by asphyxia and death when breathed pure, and frequently proves fatal when mixed with 20 per cent. of ordinary air. Accidents from this cause are by no means uncommon in large wineries. In our wine cellars (Portugal), where there is often only too much ventilation, where people are continually coming in and out, such accidents are hardly ever met with. Still, under any circumstances, it is no difficult matter to dispose of any cause for serious danger.

Carbonic acid gas, which, through being colorless, we do not notice in the air, is much heavier than ordinary air, and on this account when not mixed through the air will occupy the lowest stratum, forming a sort of bed of greater or lesser thickness. This circumstance accounts for how it has sometimes happened that men who have gone into those huge fermenting vessels have felt no inconvenience so long as their heads were above the top, but when they stooped they fell asphyxiated, and would have died if assistance had not been on hand. The same might happen outside of the fermenting vats, and on the floor of the room, when first going in, in the morning, when the fermentation had been very brisk during the night, with the

doors and windows all closed. More dangerous still would be the case if the fermenting
room were a cellar, or a room below the level
of the surface. These underground fermenting
rooms should be absolutely condemned; for,
although it may be possible to get rid of carbonic acid gas by means of lime-water, yet it
causes trouble and is very liable to be neglect-
ed. When the flow is on a dead level with or
raised above the ground, there is no difficulty in
getting rid of the carbonic acid gas, for then it
can be easily swept out, or it may fall by its
own weight. When men have to go into those
great vessels and have to work in them after
fermentation, it is always desirable to be very
careful. There is no difficulty about knowing
whether or not it is safe for men to go into
those large vats, because if a lighted taper will
not burn the air is unfit for human breathing.
In this case some means are needed to get air
in and carbonic acid out.

The success in making good wine depends as
much on the good kinds of grapes and their perfect ripeness as upon a good and regular fermentation. With the best of grapes and a badly managed fermentation it more frequently
happens that the wine will prove very indiffer-
ent than with only middling grapes well fermented. By carefully observing all the precautions

which I have pointed out, watching assiduously
the progress of fermentation, and vatting at the
time and with the care which I have indicated,
we may calculate with all but certainty on having as good and as sound a wine as we have any
right to expect from the quality of the grapes,
and the character of the season, and the state
of weather during the vintage.

The vintage made with dispatch, regularity
and neatness—that is, so as to fill one fermenting vat each day; with separation or mixing of
the kinds of grapes in conformity with the kind
of wine intended to be made; with the most
scrupulous exclusion of every faulty berry,
whether rotten, shriveled or green; with must
perfectly prepared by crushing and aerating;
with previous and absolute cleanliness of every
vessel or implement which is to come in contact
with the must; with a temperature in the fermenting room uniformly somewhat above 60°
F.; protection of the must from contact with
the air as soon as tumultuous fermentation has
set in; care in keeping the fermentation regular, with the pomace (skins, etc.) below the
surface which is to be preferred to forcing it
down daily; and finally, vigilant attention to
the progress of the fermentation with the view
to drawing of the wine when the tumultuous
fermentation has ceased, and by this means

avoiding the inconveniences of a prolonged infusion—such, generally speaking, are the means to be adopted when we endeavor to make good wine.

It is not always an easy matter to determine exactly the best moment for drawing off the wine into casks. The hydrometer, or any other instrument that we use, can give no more than approximate indications. Taste, smell and general appearance form the best criterion for men accustomed to judge of a particular variety of wine. In the majority of cases, but not in all, the disappearance of the sugar, the smell and taste of wine, a certain astringency and roughness, the color which the wine ought to have, and the formation of small bubbles on the sides of a white porcelain or silver cup, when shaken briskly in it, are almost certain signs that the wine is made. By this, however, I do not mean that it is completely finished, nor that the sugar has all been broken up; it is not desirable that it should be, because the slow after fermentation in the cask has its own advantages.

Since the new wine is always turbid, it is useful to filter some through filter-paper and view it in a glass, the better to observe its color and some others of its qualities.

The above indications of the tumultuous fermentation having ceased, are accompanied by another, which by many is considered alone sufficient—the sinking 'of the cap. During tumultuous fermentation it was raised above the original level, but as soon as this movement ceases and the evolution of gas slackens, and the temperature, raised by chemical reactions, falls, it begins to contract and sink towards its original level.

In alcoholic fermentation there are two distinct periods to be noticed: the *tumultuous*, which, as the word suggests, is accompanied with violent movement; and the *slow*, which goes on as all the materials requisite for it are present—that is, until all the sugar shall have disappeared, or until the external temperature has become insufficient to sustain its activity. The first, for red wine, has to be carried on in presence of the solid parts of the grapes. which supply it with nourishment, and yield the coloring matter, tannin and salts, necessary for making it. But when these effects have once been obtained, it always does harm to leave the wine in contact with the pomace, above all if the air can act upon it. Hence it is clear that it should be drawn off as soon as ever the tumultuous fermentation is over. No particular length of time can be assigned for this fermentation, but in general it does not exceed from four to six days.

When we have decided to draw off the wine,
it is usual, where open fermenting vessels are
used, to submerge the cap and stir it through
the wine to strengthen it, or rather to equalize
it, for it is known that what is lying in the middle is fuller in body, and possesses more color
and spirit than the rest. After a few hours,
when the pomace has again begun to separate
from the wine, the drawing out is proceeded
with. But before the mixing takes place we
should be perfectly certain that no bad fermentation has taken place—that no mould has
formed upon it. Should any such appear it
should be removed with scrupulous care.

The wine is usually drawn from the tank or
fermenting vessel through a tap fixed near the
bottom; but as it is liable to be choked by
skins, and there should be placed securely
at the back a strainer of some kind, such as a
wicker basket, before the vessel was filled, as
it hardly can be fixed afterwards.

In the inst·nce of covered fermenting tuns,
and with perforated lids to keep the pomace
under the wine, the lid and covering should be
removed some time previous to opening the tap,
taking care that the man who does the work
shall not be put in danger of his life by carbonic acid gas.

If the arrangement of the establishment is

such as I have described in the preceding chapter, nothing can be easier than to send the wine through pipes or tubing into the casks prepared to receive it; but in any case it will prove handy to run the wine first into a sufficiently deep, but not very large vat, to allow any gross impurity which may have come through, to be deposited. When the fermenting vessels stand on the same level as the cellar, the above method cannot be employed, and recourse must be had to carrying the wine in buckets, or using a suitable pump. An apparatus has been invented by M. C. Laburthe, of Mont-de-Marsan, very convenient for this operation, the principle of it being pressure of air. In some countries huge syphons are used for drawing off, but this method is far inferior to the tap.

Pressing of the pomace takes place directly the fermenting vat is empty, to obtain, before injury can take place, the remainder of the wine contained in it. In going about this work, the first thing to be attended to is getting rid of the body of carbonic acid gas contained in the vessel, because it might injure or kill the workmen. This is done by agitating a green bough, or clothes, sacks, etc., in the interior of the vessel, so as to let ordinary air in and expel the carbonic acid gas. When that has been done, the workmen empties out the pomace, which is then carried to the press.

When the wine making is done in large tuns with man holes, the drawing off is done in the same manner; and then it is even easier to extract the pomace, for it can be done with a rake through the man-hole; and afterwards, when a man has to go in and wash out the residue, care should be taken that the gas is all out, and the tun full of ordinary air, or he may die from asphyxia.

Red wine being necessarily always fermented on the skins, there will always be two qualities, which the French call *vin de goutte* (that which is drawn off through the tap), and *vin-de-presse* (that which is forced out of the pomace by pressing it).

For this purpose there are many different kinds of presses in use in different wine countries. In Portugal the kind in almost universal use is what is called *vara e parafuso*—beam and screw—a very primitive sort of a machine, occupying much room, very heavy, and not very effective, and affords no advantage beyond the simplicity of construction, and facility of repair. I would advise all who need a new press to obtain one of the modern make, of which plenty of models or plans may be seen, while their price is not high. Whatever be the press used, as all the wine contained in the pomace is not squeezed out by one pressing, and as it is always necessary to cut and mix the pomace three or four times, the wine which ru s out on each of those pressings is kept separate, and marked as first, second, and third, for there is

difference in their composition, which contains
different quantities of the materials yielded by
the solid portions of the grapes. The first pressing generally contains the most alcohol, tannin
and coloring matters in solution; the others
more extractive matters.

Some viniculturists are opposed to the mixing
of any pressings with the wine, as they consider
them to interfere with its fineness, softness and
sweetness.   Others again contend that the
pressing, especially the first, improve and give
endurance, color and consistency to the bulk of
the wine, whilst without the addition it would
remain too soft and not keep well. This view
is favored by the best authorities, and has the
recommendation of being in general use in the
best wine districts.

It is true that without them the wine ripens
sooner, and becomes fit for use; but then there
is the risk of its not keeping, and its early decay. On the other hand, the result of mixing
the wine with one-fifth or one-sixth of the
pressings is that at first it is rougher and hard-
er, but it always keeps better, and with time
loses this harshness, which is disagreeable only
while it is young. The good properties of wine
treated in this manner are ascribed to the larger
amount of tannin in solution, principally that
derived from the seeds, because it possesses

the property of precipitating the ferment, which may still be in the wine, and may become the cause of its deterioration. Under these circumstances it is beyond all doubt advantageous to add a portion of the pressings to the wine.

In order that the mixing may be properly done, it is necessary that it be uniform. If all the wine of one fermenting be put into one vat, it should be of such capacity as to hold both the wine and the pressings, especially the first pressings; not forgetting that it is always useful to set aside a certain quantity of the same wine in a small cask, to be used for filling up, an operation which it is not proper to perform with wine of a different quality. When the mixing has to be done in a number of casks great care must be taken to preserve the same proportions in each, so that the bulk may prove uniform. In this case the casks are filled three parts full, reserving one third for pressings.

In Burgundy and in the Medoc the wines are matured in barrels of 280 litres capacity (60 Imperial gallons nearly), and the wine is distributed so evenly through them that each one receives an equal quantity of wine from the top, middle and bottom of the vat. Of the first wine which runs out of the tap and which corresponds to the lowest part of the vat, they put equal measures into each hogshead, one

4

after another, and so of all the rest of the contents of the vat; and lastly the same with the pressings till they are all full.

Obviously the same result can be obtained, and even with greater certainty that it will be uniform, by making the mixture in a sufficiently capacious vat, stirring the materials thoroughly, and after a few hours of rest to allow the grosser parts to fall to the bottom, draw it at once into the casks in which it is to remain till the time for racking.

What I have just said about mixing the pressings with the wine, applies in a general way to the making of all table red wines, or pure wines of food; still, as this question is surrounded with many conditions, such as the nature of the kind of grapes used, their ripeness, the state of the stalks, etc. (*engaço*), and the character of the wine we wish to obtain, it is plain that the viniculturist should attend to all these points, and seek in experience for the best plan of conducting his operations.

Throughout this exposition of the processes of wine making, I have again and again insisted on the necessity of absolute cleanliness and neatness of all vessels, utensils, things and persons coming in contact with the wine. The majority of our wine makers unwisely undervalue the precautions necessary for making wine that will keep and mature well; and I cannot help insisting on this point again even at the risk of wearying the reader.

The general precept is very simple—whether in bringing in the grapes, the making of the wine, or in its after manipulations and care, never allow to be used any utensils, instruments or vessels which have not been washed scrupulously clean. The baskets of the pickers, the boxes in which the fruit is carried, the tubs and carts in which they are conveyed to the press room, the tanks or benches where the crushing takes place, the presses, buckets, and other vessels used in carrying the wine, the tubing used for passing the must into the vats, the hoppers and funnels, all the machinery, the tuns, pipes, and all the apparatus of what kind soever are comprised under the general recommendation—they all need washing before being used; and as soon as the wine making is over, again that they may be clean and sweet for the next vintage.. The greatest of all care and vigilance are required about the fermenting vats and vessels, because through being in contact with the wine, they may so easily communicate to it any dangerous matter with which they are foul, from which may arise, not alone a bad taste, but the ruin of the wine in the future.

New wooden vessels of all kinds ought to be

well scalded with boiling water, and afterwards
washed in cold water, before being used.
Utensils which have been in use require to be
thoroughly examined on the inside if they have
any bad smell, and if they impart a bad taste
to the water when washed; if the staves show
stains in the wood, because they may be the
earliest indications of rotting, or any other defect, in which case they should be coopered
and planed, and afterward washed with water
containing 1-10 of sulphuric acid, and rolled
about so as to bring it in contact with every bit
of the inside; and afterward washed with cold
water and a brush until the last trace of the
acid is removed. Some employ with much advantage freshly prepared milk of lime, always
however finishing with a wash of cold water,
and best of all—for a final purifying, a *rinse with
alcohol.* After being well washed, the vessel
should be dried with a clean cloth, and if it be
not to be used right there, a match should be
burned in it and the bung driven home. This
burning of a sulphur match in the empty casks
should be repeated every six months, for nothing is more efficacious in preserving casks from
mildew.

The employment of good spirit to wash the
casks in which fine wines are to be kept is very
useful. In some countries, for instance in Bur-

gundy, they use wine of good quality, and this is sensible, because it cannot communicate any foreign taste or smell.

Writers on the treatment of wines have given a host of receipts for cleansing wine casks for maturing wine, but I see no need of reproducing them in this place, convinced as I am that what I have said already is ample for this part of my work, if it only be faithfully performed.

Having now arrived at the point of putting the wine into casks, let us see what care is needful for the complete organization of the wine, and to insure its ability to keep.

CARE NEEDED BY THE WINE AFTER IT HAS BEEN PUT IN CASK.

Whether it be in casks of great capacity, or of medium size, or in pipes or barrels that the new wine is put for keeping, it is absolutely necessary to leave the bung-hole open to admit of the escape of gas generated by fermentation, and which for some time continues in the liquid.

When drawn off from the lees, the wine always contains a certain portion of sugar and ferment, and what comes from the press has still more. The aeration which the wine undergoes in the process of vatting and in the press, furnishes a sufficient amount of oxygen to impart new life to the ferment, which soon becomes apparent by its greater or less activity, but which gradually diminishes till the bubbles of gas are scarcely perceptible. As long as this secondary fermentation continues, it is only necessary to cover the bung-hole with paper, or, better still, two thicknesses of muslin, to keep dust and flies out. The body of carbonic acid gas floating on the surface is enough to preserve it from contact with the oxygen of the air. In some parts, for sake of greater security, a simple tin tube, which acts as a hydraulic valve, in the form of an inverted letter ∽, is adapted to the bung-hole, with some water in the curvature, or cotton wool.

When there is no longer any sign of fermentation, whether because all the sugar has been broken up, or on account of the temperature having become too low for fermentation to continue, the bung ought to be driven home to exclude all contact with air. This usually happens during November. Just about that time, the wine, heretofore more or less turbid, begins to clear itself and become transparent. They then say "the wine is made;" yet from this point onward the need of care to preserve it and perfect its organization is not less urgent.

The first condition to be attended to, is the complete exclusion of air, in order to prevent the formation on the surface of those vegetable growths or *flowers* which precede or accompany the acetic fermentation, and other changes which injuriously affect it. In order to effect this exclusion of air, it is indispensable to keep the vessels completely full to the bung. The vigilance should know no bounds on this point, most of all during the first period of its life, for either indifference or carelessness may occasion serious injury which later on will show itself, and when too late to be remedied.

In all kinds of wooden vessels, two causes concur in diminishing the volume of wine contained in them, and by consequence a vacant space which the air soon fills. The first is contraction of the liquid by cold, and the second evaporation of the liquid through the pores of the wood, and the interstices of the bung. This last cause goes on all the time, and hence the need of filling up at very short intervals. During the first months it is necessary to fill up at least once a week; later once in a fortnight, and lastly once a month, but never less. After every filling, the bung should be wrapped in a clean cloth dipped in the wine, or in good brandy. The wine used to fill up the casks should be identically the same as that which they contain, scrupulously preserved from any alteration, in fact, beyond all suspicion. For this purpose a sufficient quantity of the same wine should be stored in small vessels of some convenient kind and size; but, for the same reason that compels us to keep the big vessels full, we have to take care that that used for filling up is not allowed to lie in contact with air. So the vintner needs to make his calculations how best to comply with these general instructions.

In proportion to the number and capacity of the keeping casks, quantities should be provided, each just sufficient for one filling up. Where the quantity is small it is best to have the wine in bottles to avoid loss and prevent injury. This is the practice in countries where fine wine is made and matured, as in Burgundy and the Medoc, where it is kept in barrels of only 228 litres.

A very simple and effective method of keeping the casks full is to drop in perfectly clean quartz pebbles.

It is easy to understand the reason for all these precautions; since if the wine used for filling up were not of the same quality, or contained any principle of alteration, it is clear the disease would be communicated to the whole.

Later on when the wine has been put into pipes or smaller casks, and when it is convenient to pile them one upon another, and it becomes difficult to keep them full, the proper course is to place them with the bung on one side, so that it may be kept moist and exclude air. As to the evaporation, which will still go on through the pores of the wood, whatever space is formed will be filled by vapors of the wine, which will keep out air.

The aim and object of the above precautions is to preserve the mature genuine first-class red wines; for those are the most liable to de-

4ᴀ

teriorate. The action of air, aided by a sufficiently high temperature, may originate secondary fermentations, and internal disturbances at
any time, until the wine is perfectly formed and
freed from all foreign and prejudicial matters.
For the keeping of such wines until they are
ripe for bottling, it is most desirable that the
temperature of the keeping cellars should not
exceed 10° centigrade, or 50° Fahrenheit.

RACKING AND CLARIFYING.

Rackings and clarifying, with albuminous or other materials in place of it, are operations called for in order to remove the lees and solid matters which might interfere with its keeping qualities. Both of those operations belong to the class of attentions which wine demands during its earliest periods, after being placed in cask, and they contribute much towards its completion.

When the wine is drawn into casks it is always muddy, and continues so as long as fermentation continues. But when it has quite ceased, about the beginning of the winter colds, it becomes limpid and transparent, and deposits whatever solid matter it had held in solution. This is what we call lees. These lees contain, among other things, the residue of the ferment, now inactive; minute fibres from the pulp of the berries; bi-tartarate of potassa, which becomes insoluble as the alcohol is formed in the liquid. In the lees may also be found the germs of organic life, which under favorable conditions may give rise to those several kinds of fermentations which cause the diseases of wine.

As long as the temperature remains low the germs lie dormant, and the presence of the lees does the wine no harm, but as soon as it rises to a degree of warmth favorable to fermentation they rise in the wine, which again becomes turbid, "kicks up," and is often hopelessly lost. Common sense points straight to the removal of all danger from the above source, as soon as ever the conditions are favorable. Racking alone can effect it.

Sometimes the wine does not clear early, and now and then it remains turbid, holding in suspension permanently solid foreign matters. In these cases it is necessary to force such matters to deposit. The means usually had recourse to for this purpose and found to be the safest and most efficient, is clarifying with isinglass, or gelatine, or white of eggs, or other substance having similar properties, which, upon being introduced into the wine and uniformly distributed through it, becomes more or less solidified by the action of the alcohol and tannin, forming as it were a net, which, as it sinks toward the bottom, drags the suspended matters with it, leaving the liquid clear and bright. In theory this is "clarification by albuminous matters." Let us now see how and when each of the above substances will be found best to use, when we have to employ them. In most cases, and principally in the instance of red wines, which is the chief matter of which I am at present treating, racking is indispensable, and should be done as soon as the wine has deposited its lees. As this occurs during the cold weather of winter, and as with warm weather of spring will start fermentation, the racking ought to be all over before it sets in. March is the month when it most usually is done, for then the weather is cold, clear and serene.

Necessary as it is to rack, a moment's consideration will suffice to show that neither the exact time when it should take place, nor how often it should be repeated. The nature of the wine, the conditions under which it was made, and the state of the weather—all may modify the periods for racking.

Wines loaded with much ferment and foreign matters, if the weather be fine in early winter—November or December—should then have a first racking. This racking, however, has not to prevent a second before the opening of spring, because this is indispensable in the case of *all* wines in order to protect them against the first warm weather of spring, as also at the time when the vine breaks bud, for at that time they almost always show signs of internal commotion.

The same disturbance appears to be renewed

when the vine is in bloom, and again when the
grapes begin to change color in August. In
very hot years, and in exceptional cases, it may
be desirable to rack in the latter end of June,
chiefly if the work was badly done in March, or
not at all.

Racking becomes all the more indispensable
when the wines are poor and defectively made.
All the ordinary ones, especially in bad years,
demand at least two rackings, the first one being preceded by a fining with isinglass or other
matter, which will be treated of later on.

Fine alcoholic wines made with very ripe
grapes, with much sugar and little ferment,
and which consequently deposit little lees, may
need only one racking, at the end of winter.

The operation of which we are treating, in
order to be beneficial and safe, requires a certain number of precautions and cares to be attended to with all vigilance.

In the first place, it is proper that the wine
should be clear and that lees should have completely settled to the bottom. When this has
not taken place by the time when we judge the
racking ought to take place, then we must re-
cur to previous clarification.

For racking we ought to choose a time when
the weather is clear, dry and cold, with a north
wind, if possible; because then the whole body

of the wine is tranquil, the air pure, and there is little chance of hurtful germs being introduced into the wine.   On the other hand, rainy, blustery, stormy weather, and the winds from the south (in the northern parts of Portugal) are not favorable to a good racking.   Experts affirm that the best time for it is early in the morning.

It is quite essential that as far as possible no air should be allowed to come in contact with the wine during racking, particularly in the case of delicate wines, and of no great body. The method commonly in use in Portugal has been to insert a tap in the cask a little higher than where it is supposed the lees have settled; draw the wine into buckets and pass it into another cask previously sulphured.   As soon as ever the wine comes over turbid, stop; and put the thick wine and lees into a vessel apart. Part of the turbid wine may be clarified afterwards, and the lees used for distillation.

In doing the work this way, it is impossible to prevent contact with the air, and the loss of some portion of the brandy and aromatic elements; and on this account many fear to rack wines which are naturally weak, or which are unusually delicate.   The danger, however, will be greater from leaving them on the lees, than from exposing them to the loss of some alcohol

and perfume; which, however, need not be lost
if the method I am going to describe be adopted—one which is in general use in the most
celebrated wine districts of France. The cask
which has to receive the clean wine, is placed
close to the one from which we are about to
rack. In a hole made in this latter, we fix a
straight tap, to which is fastened a flexible tube
of leather or rubber. To the other end of it,
another straight tap is secured, which is fixed
in a corresponding hole of the empty cask.
When a vent-hole has been made in each, to allow the passage of air, and both of the taps have
been opened, the wine will run into the empty
cask till it reaches the level in each. In order
to force the rest of the wine into the clean cask,
and out of the other, they fix securely in the
bung-hole a special kind of bellows. The air,
compressed in the cask, acting on the surface
of the liquid, forces it to enter the other cask.
When they hear a sort of whistling noise in the
tube, indicating that air is entering, both of
the taps are closed, and withdrawn. The rest
of the wine and lees are drawn off into a bucket. By this process, which in practice is sim-
pler then it appears in the description, especially as to the tubes and bellows, the racking is
done without difficulty, without allowing the
air to exert any notable influence on the liquor;

and effectually preventing the loss of spirit and perfume.

Racking can also be done with a syphon or a pump, but with less advantage and more risk than by the method I have explained, especially when the casks are small or middle sized. When racking has to be done from one great vessel to another of equal size, the bellows cannot be used, but an appliance devised by M. Laburthe can. It is, of course, an easy matter to rack off wine from a great vat or tun, into casks. We have to take notice however, that wines of high quality gain much by being kept in the casks into which they were first drawn. To this end, as soon as the cask is emptied, carefully washed and sulphured, the wine should be returned to it, the same precautions and appliances being used as in the racking. In Burgundy and in the Medoc, wines of high price and estimation are always drawn from the fermenting vats into new barrels of 228 litres, so that when the racking and clarifying have been done, they may be returned into them. In them they are sold, and the price of the cask is added to that of the wine.

ON CLARIFYING.

When wine by simple racking becomes limpid, transparent and "candle-bright," there is no need of clarifying; but this rarely happens, and the wines, even after being well racked and at a proper season, fall still short of being quite clear and bright, on account of minute matters still held in suspension, and which must be got out, because sooner or later they may injure the wine.

It is just as impossible to state any fixed times when wine should be clarified as when it should be racked. Common sense must be used about the condition and age of the wine, the state of the weather and other circumstances. At any time after the racking, if the wine be not bright, it may be clarified if the weather is cool and clear, and, under any circumstances, when the wine has to be either exported or bottled, it ought to be clarified. During their first year all wines require at least one clarifying, and those made from bad grapes or in a bad season, that have an excess of ferment and clear with difficulty, demand clarification absolutely. But, above all others, wines which have a natural tendency to "kick about," as we commonly say, require rigorous clarification after their first racking.

Artificial clarifying accomplishes, as far as possible, the same effect as would be if the wine were filtered through paper. But as this cannot be done without ruin to the wine, we prepare a substance of some kind which will quietly fall through the wine and accomplish the same purpose. Many different substances may be employed for this purpose. The more generally used, however, are whites of eggs, consisting mainly of albuminous matters; isinglass or albumen, of which there are several varieties; the blood of bullocks or sheep, which contain matters that coagulate, such as albumen and fibrin; fresh milk, which contains the curd also coagulable; and many artificially prepared substances, prepared expressly for clarifying wine—all of which have more or less the properties just mentioned.

It is by no means a matter of indifference which of these substances is to be employed, for the one selected may either clarify well or not at all, or yet, what is still worse, leave behind it injurious matters.

I have always found it advantageous, just before racking in March, to determine the alcoholic strength, and again after it; so that if there have been any loss, it may be restored by the addition of some good brandy, or enough to leave it very slightly in excess.

What I have already said about the advantage and even necessity of separating the clean wine from the lees is always applicable to pure, genuine wines for the table. This refers mainly to red wines, but there are special cases where a wine may be benefited by being for a time left to lie on the lees, at least as long as the processes of its formation are progressing. This has special application in the instance of white wines, for they are not fermented on the stalks and skins; and we have found in many of our provinces that rolling the casks repeatedly, and mixing lees and wine thoroughly, is good practice. It is the weak, thin white wines, and even the more spirituous, which, from different reasons, are benefited by this mixing through them of the lees during their first year. I have alluded to the advantage, and, in most cases, the necessity of artificial clarification of new wines; let us now inquire into the conditions of how and when it should be done.

For clarifying red wines, nothing appears to surpass the whites of new-laid eggs; for white wines, pure white gelatine, or isinglass is preferred.

Albumen, which forms an essential part of the

whites of eggs, and which we find in bullock's blood, is a substance which without alteration of its composition, can exist under two forms. It may be either soluble or insoluble. In fresh eggs it is in the soluble form (can be mixed perfectly with water). When heated to 75 C., or about 180 F., it becomes fixed, and can no longer be dissolved in water. But it is not alone heat which coagulates albumen; alcohol, tannin, strong free acids, and other chemical substances coagulate albumen, whether pure or dissolved out in water.

Gelatine, which forms the different sorts of the glue of commerce, is not very *soluble* in cold water. It softens and swells up, but it dissolves readily in warm water, and forms a jelly. The solution is coagulated by alcohol, which throws it down. Tannin forms an insoluble compound with it, and therefore precipitates it from its solutions.

In view of these properties of albumen and gelatine, it is not difficult to understand how they act in the process of clarifying wine. It holds in solution alcohol and tannin. Where solutions of albumen or gelatine are mixed in matters like wine, containing alcohol and tannin, they form a kind of membraneous net, which, little by little, contracts and gathers up whatever matters they find in suspension, and

by their united weight, drag them to the bottom
of the vessel, and of course, after a while, leaving the liquid clean and bright.

As already said, the white of eggs is the most
convenient for clarifying red wines generally.
It is the only substance which should be used
in clarifying red wines of high character, for
about its purity there can be no question,
neither does it introduce any dangerous foreign
substance. Ordinarily the whites of two or
three eggs suffice for one hecolitre of wine
(about 26 gallons). The whites of the eggs
should be quite free from any yolk; then in a
suitable dish or other vessel, beaten up with
some water and clean salt, into a froth. Then
after drawing off some quarts of the wine to be
clarified, into a bucket, the eggs are first added
to it, and the whole thoroughly incorporated
with a piece of clean wood, flattened at the end.
and a cross piece attached to the end, so as to
make the mixture as uniform as possible. This
mixture is then poured into the cask of wine to
be clarified, stirred well in it with a suitable
wooden implement, and left at rest for a number of days, more or less, in accordance with
the size of the cask. If the clarifying has been
done in a "pipe," at the end of eight or ten
days, the sediment will have fallen to the bottom, and the time has come for the second

racking. The salt employed has the advantage of favoring the solubility, and as a consequence the division and distribution of the albumen through the body of the liquid, and produces a more gentle and effective coagulation of the suspended impurities: for itself is insoluble in wine, while on the other hand it acts as a preventive against alteration, corruption, or fermentation of the deposited lees.

From what has been said concerning the manner in which the clarification is effected, one might conclude that all that is necessary would be to mix the whites of the eggs, prepared as above, with the wine at the top of the cask or tun; and this would appear to be an advantage in the case of very large vessels.

[N. B. This is theory, and needs practical confirmation, especially as to the shape of the casks or tuns to be clarified.— *Translator.*]

Clarification with gelatine, or other analagous substances met with in commerce, is certainly very effective; but before speaking of the methods of using it, it seems desirable to say a few words concerning considerations which should guide us in their employment.

The essential condition of a good fermentation is the formation of that kind of a membraneous net, which contracting itself in the liquid, can lay hold on and carry down the impurities and germs of fermentation suspended in it. Now all sorts of gelatine are not equally efficacious in producing these effects, and not a few of them, on account of being badly made, dirty, or not fresh, may impart dangerous matters to the wine.

The gelatine obtained by prolonged boiling of animal skins, and other animal tissues, which furnish the white gelatine of the painters, does not satisfy the requirements just mentioned, and is useless for clarifying wine. Still this same, when once perfectly dried, constitutes the strong gelatine, when free from bad taste and smell, may be used, and is frequently used with advantage in clarifying wines containing an excess of tannin. Since it precipitates tannin, it softens the wine, as well as clarifies it.

The quantity commonly employed is 20 grams (about 8 ounces) of the dry gelatine to about 26 gallons of wine; but of course this is a matter to be left to the judgment of the operator, since the wine may be very muddy, or nearly bright. There are varieties of gelatine so white and fine, that as much as 35 grams per hectolitre might safely be used; and with such gelatine even fine wines may be clarified. The process employed differs little from that employed in the use of albumen. The gelatine broken into small bits, is put to soak in a little water; after a time a slow heat is applied; then it is thoroughly mixed with a small but sufficient quantity of wine, drawn from the cask, and then this mixture is put into the wine to be cleaned, when the cask is to be closed and left to rest.

The variety known in commerce as the fish gelatine (isinglass), obtained from the air bladders of certain kinds of fish, notably the sturgeon, is far preferable to the gelatines, however pure they may be, but above all in clarifying delicate white wines containing but little tannin.

This substance does the clarifying independently of any action on tannin, keeping its membraneous form, even when very much diluted, and acting mechanically; being contracted by the alcohol, and falling through the wine without robbing it of anything which constitutes its body. It is however used cautiously, not more than 20 grams to a pipe, or about 4 grams to the hectolitre. It must be finely cut, soaked with a little water for a few hours, then well broken up with the hands, or in a collander, diluted with some wine in a bottle or flask, and and then used for clarifying in the ordinary way.

Clarifying with the blood of bullocks or sheep, does its work with energy, but it is nev-

5

er employed except to clarify very ordinary
wines, because they nearly always leave behind
them a sickly taste, and throw down a heavy
deposit.   When it is used at all, it should be
sparingly, and always with quite fresh blood.

The powders and artificial preparations for
clarifying wines, whose composition is held a
secret, offered in commerce, have as their principal component or base, dried albumen obtain-
from the serum of animal blood.   They offer
no advantage over the means already spoken
about which could counterbalance the risk of
deception.

Over and above the animal materials, there
has been a mineral substance pointed out, which
can be easily obtained, and in sufficient purity
for clarifying wines without risk of any kind.
This is alumina in the gelatinous state. which
is prepared by decomposing a solution of common alum by an alkali (soda. potassa or ammo-
nia), and thoroughly washing with pure water
the gelatinous precipitate formed.   This gelatinous alumina is insoluble both in water and
wine, but it can be easily mixed through the
body of liquid by agitation, when it will fall to
the bottom and draw the impurities along with
it.   It has, however, the property of absorbing
coloring matters, with which it forms lakes; so
its employment in red wine, at least, will al-

ways be attended with loss of color. Since it also forms an insoluble compound with tannin, it will necessarily diminish the roughness of the wine. As yet it has not been adopted in general practice, and the most I can say for it at present is that it should be used only in an experimental way and with much precaution.

We observe sometimes how wine will not clarify even when the best means have been employed. Now, the cause might be a movement of fermentation going on in the liquid, or a deficiency of those substances which are indispensable for precipitating the albumen or gelatine—that is, a want of alcohol and of tannin. In either case we must not go on adding more albumen or gelatine. Since clarification requires the liquid to be perfectly tranquil, it is clear that if any fermentation is going on the clarifying materials have no chance. In such a case it is best to rack the wine into a well sulphured cask, and wait till all signs of fermenting have disappeared, selecting for the purpose cool, clear weather. This supposes the fermentation to be alcoholic, for if not that, then the wine is diseased, and other means have to be availed of, concerning which I intend to treat later on.

If the clarifying matter is ineffective on account of the weakness of the wine, the best thing to be done is to rack it as before said, and when clarifying is again undertaken, at the very time of doing it, to give the wine a fair dose of brandy and tannin, proportionate to the amount of clarifying matter used, and for this purpose it would be well to make an experiment or two in a graduated glass or tube. The best solution of tannin, or at any rate the most easily prepared for this purpose, is got by soaking grape seeds in good brandy, a quantity of which should be always at hand in the cellar, for it comes in useful in other cases to be treated of later.

To resume. Clarifying having for its object to remove out of wine impurities which might injure it, especially the germs of fermentation, which might alter its constitution, is a very necessary operation, and above all indispensable for the preservation of fine delicate wines. There is nothing either difficult or expensive about it. Whites of eggs, gelatine and isinglass are the substances ordinarily employed, with the best results. The two first are used to clarify red wines, the last for delicate white wines. The clarifying should be done in bright serene weather, and, when possible, in cold weather after racking. Immediately the wine is clarified, and the matters well settled in the bottom of the cask, it is proper to rack it again. During all the time, while clarifying is in progress, the wine should remain perfectly still and undisturbed, not alone as to the liquid in the cask, but any blows or shocks on the outside.

With suitable rackings, with opportune and rigorous clarifying, and with carefully attending to filling up, the wine will be in good condition, will keep on improving, and acquiring the best qualities that its nature is capable of developing as time passes.

SULPHURING.

Already on several previous occasions I have pointed out the necessity of purifying casks, etc., by burning a sulphur match in them before filling wine into them, with special reference to the occasions of racking. Sulphuring is a powerful help in maintaining the health and keeping properties of wines, and on that account it may be desirable here to enter somewhat into details.

Sulphuring by burning a match dipped in sulphur, or sulphur alone, it matters not which, may be useful under two points of view, either for sweetening the casks, since it will absorb oxygen and kill fungoid germs, or to arrest alcoholic fermentation for a longer or shorter time, or suspend fermentation of the sugar. The first case is the most general, for it is applicable to all wines whatsoever; the second finds its use in making certain kinds of white wine, in which the aim is to preserve an excess of sweetness, by its protecting the sugar against the ferment.

The theory of its operations is of its own nature very simple. When sulphur is burned in presence of oxygen the two combine and form a colorless gas, which cannot be breathed, and which has a suffocating smell. We call it sulphurous acid. This gas is readily soluble in water and in wine, and after being formed it possesses the power in the presence of water or moisture to absorb still more oxygen, becoming converted into sulphuric acid, commonly called "oil of vitriol." Hence we see that when sulphur is burned in a cask or any closed vessel or place, it first lays hold upon the oxygen of the air, leaving sulphurous acid in its place. In air so deprived of oxygen, and in presence of sulphurous acid, it is impossible for those lower forms of organisms to live, such as the mycoderms and cryptogamic vegetations, which are vulgarly called mildew, mould, and ferments, properly so called. Even in solution sulphurous gas kills them and stops all fermentation. If, by absorbing more oxygen, it becomes sulphuric acid, this exerts a corrosive action on those minute bodies. When a cask has been emptied and not washed for a few days, its inside will be found covered with mildew, imparting to it a bad smell, starting in the direction of rotting the wood itself, and rendering it unfit to hold wine, for it would certainly be ruined. The way to prevent this is to sulphur as soon as the cask is emptied of lees, and has been well washed.

For sulphuring we prepare matches, which
are strips of linen, cotton or paper, 25 centimetres long, and 5 centimetres broad, more or less,
which we dip in melted sulphur, and hang up
to drain. When we use them we attach a length
of wire, light the match and lower it into about
the middle of the cask. While it continues to
burn the bung hole should be closed, and the
wire secured on the outside. A bung with wire
already attached is a convenience. In withdrawing it, care should be taken not to drop
the ashes, as they might possibly affect the taste
of the wine. A portion of a match, of 10 or 12
superficial square centimetres, is enough to
sulphur a pipe. In the case of sulphuring
large vats, tuns, etc., burn the sulphur in an
iron, or porcelain capsule, properly secured to
a wire, which may be once or twice raised up
and down.

When we go to rack, no time should be lost
when the sulphuring is finished, in pouring the
wine in while all the fumes are there; otherwise
the benefit is not only lost, but it might prove
injurious; because the sulphurous acid, coming
in contact with the damp sides of the cask,
would be converted into sulphuric acid, and
impart a bad taste to the wine afterwards.

When however we sulphur empty casks to
keep them sweet, we drive the bung tight while
all the fumes are in them, and leave it so till

needed for use; and then before using them, they should first be well washed, to remove any taste that might have been caused by the old sulphuring.

We observe occasionally that the match will not burn inside a cask which has been used for some time. This is caused by the bad condition of the staves on the inside; because some or all are in a state of decay, or covered with mildew, or other form of change, which had absorbed the oxygen, for in air without oxygen sulphur will not burn. It is in such case necessary first to fill the cask with air by using bellows, or by any other method, the best of which would be to make the cooper take out the head and examine it.

Sulphuring becomes of very much use, whenever we have to keep a cask not quite full. In all such instances the oxygen of the air tends directly to cause change; mycoderms develop, or the flour of wine and of vinegar; generally it turns sour and is lost. Up to a certain point, this may be obviated by burning a match in the empty space, and then turning the cask so that the bung shall be covered by the wine.

When we wish to stop fermentation in wine or must by sulphuring, it is necessary to actually saturate the liquid with sulphurous acid gas, which cannot be done by burning only a match

in the cask.   First we burn a match, and while
the cask is full of the fumes, pour in a portion
of wine, close up the bung, and roll it about
till the liquor has absorbed them.   Then remove the bung, admit the air, burn more
match, and add another portion of wine, and
so on till the cask is full.   Thus the whole of
the wine or must will be saturated with sulphurous gas, all fermenting put a stop to, and the
sweetness preserved, but at first giving the disagreeable taste of sulphurous gas, which after
a while passes off.

To recapitulate, sulphuring is eminently useful for preserving both wines and casks.   When
they have to remain empty, it protects them
against mould, and other hurtful influences,
while wine in presence of it loses its power of
fermenting and remains unchanged.

Since during the burning of sulphur the oxygen is abstracted from the air, and sulphurous
gas produced, and since the presence of oxygen
is indispensable to the wine during certain periods of its formation, evidently it cannot be
made use of before the wine is made, or when
we desire to stop fermentation, to preserve a
part of their sugar.   Sulphurous gas has a very
energetic action on coloring matters, bleaching
them, and causing them to disappear.   On this
account sulphuring is used in white wine, when
it is sought to have them pale.

There are but few cases in which sulphuring is *not* useful, and they may be reduced to almost the following:

1.   When we wish to preserve for certain kinds of wine that slight effervescence, which is a characteristic of wines of Bucellas, and which is due to very feeble fermentation.

2.   When the wines are too alcoholic, and which ferment only slowly, as occurs when raisins are used.

3.   Lastly when the wine has been attacked by some particular disease, which will be treated of hereafter; when it assumes the appearance of oiliness, or fatness, in the treatment of which oxygen is needed, and therefore sulphuring would be out of place.

When speaking of sulphuring casks, I ought to make a remark or two, which perhaps the educated and practical wine treater will kindly excuse.   The remark refers to a disaster which has occurred oftener than once.   Sulphuring must never be done without first making sure that there is no brandy in the cask, nor even the vapor of it arising from previous washing, since the mixture of the vapors of alcohol and explosive air in presence of the burning match, will assuredly produce a violent explosion, which may cause great damage, not to say burn the buildings.

THE USE OF BRANDY.

The employment of brandy to keep and improve wines is so general in Portugal, and considered so indispensable, that they have adopted into general vogue the words "*beneficiar os vinhos*" (benefit the wines). In making Port wines, and some others which are made in imitation of them, and also in making liqueur wines, it is indispensable, they say, to use it very freely. It is not my present intention to enter upon any discussion of the generous wines of the Douro, and others like them, because, as I have already repeatedly said my especial object in this little work, is to treat principally of genuine, pure, nourishing table wines, like French Clarets.

The question arises, is the use of spirit an advantage in making wine, and is it indispensable in the making of natural food wines?

This question deserves to be treated at considerable length, for there is a generally prevailing opinion that our wines do not keep well, nor endure voyages well without additional brandy. But on the other hand, wines loaded with alcohol have become objectionable to the public taste in England itself, which has been the chief consumer of our excessively generous wines of the Douro. So, while there, the consumption of French wines is rapidly increasing, with an alcoholic strength far beneath that which we have been sending to that market habitually, our importations, if they have not fallen off considerably, remain stationary, and bear no relation to the increasing consumption of wine in England.

Without denying the utility and advantage of brandy in very moderate quantity in ordinary wines of consumption, table wines, I am convinced by individual experience, that it is by no means indispensable for the keeping of such wines, when they have been made and kept with all the precautions and cares which I have pointed out, unless in exceptional and purely accidental cases.

The southern provinces, where the grapes arrive at a maximum of maturity, and where they are the richest in sugar, and on this account would naturally produce the most spirituous wines, are exactly those in which the employment of brandy is most common, for experience has demonstrated that their wines, on account of the way they are made, will not mature and keep without a large addition of spirit. In the manufacture of the Port wines of the

Douro, the need of using brandy largely is perfectly understood. When they set about making full-bodied wines of deep color, they begin by a hard and prolonged and even violent working of the grapes, so as to reduce the berry and pulp and stalks to a high state of attenuation, charging the must in this way with all the materials needful for a perfect fermentation, not only the alcoholic, but those which may come afterwards and corrupt the wine. It therefore becomes indispensable to render inoperative those dangerous materials, and the readiest way to do is by an excessive dose of brandy, which in time proves of benefit in developing the best characters of this kind of wine, which are brought out by the action of acids, obtained not alone from the berry, but from the stalks also upon it, which results in the formation of the ethers and aroma of the wine. But while wine made by this method, with very ripe grapes acquire essentially estimable properties, and an immense power of resisting alteration, on the other hand, they lose the property of being food wines (*alimenticious*), and are, as a matter of course, too exciting and generous.

In the south of France and in Catalonia, full-bodied and very deep-colored wines are fabricated, to be mixed with the poor, thin wines of other countries, and for such purpose their excess of alcohol is an advantage, upon which the cellar men rely for their keeping. Still, those Roussillon and Catalonia wines are hardly drinkable until blended with weak ones. Their *raison d' etre* has a ground to stand on, but very different from what I am advocating.

The class of wines which our country (Portugal) for the most part is best suited to produce are natural wines — the pure juice of the grape — which is capable of being made good — nay, excellent — for every-day use at home, and fit for exportation to any country without risk of loss. That this can be done in good condition without the addition of spirit I am satisfied, for the same reasons that it is done in countries where the nature of the grapes, the soil and climate are as nearly as possible identical with those of Portugal. Still, the addition of brandy in just and moderate proportion, far from doing harm, may very frequently improve the wine, and aid in its keeping properties. Hence we should not condemn it, but keep the use of it within strict limits.

In a work published in 1864, Baron Thenard treats extensively on adding brandy to wine, in reference to France; but since, for the most part, his arguments might be applied with equal force in Portugal, I ask leave of this nobleman, whose friendship I have the honor to enjoy, to transcribe a few paragraphs from his interesting report, because they enable me to explain clearly the advantages of using brandy within reasonable limits.

" *Fins da aguardentasam* "—ends and objects aimed at by the use of brandy.   Let us now see what end we propose to accomplish by the use of brandy in wine.   Many think the only use of brandy is to increase the alcoholic strength thereby; but the end is much more complex — it is one of the most important points of this essay.   (*Exposição.*)

"In fact, beyond its direct and easily understood action, spirit exerts others more latent, but more useful, reacting beneficially on the constituents of wines, such as the acids, the salts, the color and the ferments; it arranges and equalizes in a certain way their different elements, whether by eliminating some in part, at least, when they are too abundant, or diminishing their effects; or, on the other hand, by facilitating the development of those which are deficient.

"On the importance of adding brandy to sweeten very acid wines.   We must take leave on this subject to go into particulars.   Every one knows that sour wine is disagreeable to the taste and hurtful to the stomach.   Now, without meaning to say that a wine of low spirit

strength is always acid, we may truly say that
an acid wine always contains little spirit, and
that brandy diminishes this grave defect in
proportion to the quantity added. This our
palates can tell us, but a simple experiment
demonstrates it better.

"'The wine, in point of fact, owes its acidity
on one hand to a series of free acids always
present in relatively small proportions, and on
the other hand to the presence of a salt called
cream of tartar, which has a peculiar acid taste,
and which is often so abundant that it forms a
considerable deposit on the inside of the casks.

"If we saturate water with cream of tartar
and then add a *little* brandy, the water directly
becomes turbid and deposits a white powder,
which is the cream of tartar. If, after giving
it this small quantity, we put in more brandy
the deposit will be increased in such degree
that when the alcohol attains a certain limit all
the cream of tartar will have been deposited,
and the water will be found to have lost all
acidity.

"Repeat the above experiment with clear but
very sour wine, instead of the aqueous solution
of cream of tartar. The result will be identical. Over and above this, direct experience
proves that in a wine, either naturally rich in
spirit, or aided with brandy up to 10 per cent.,

there will never appear an excess of acid. So,
just as we have said, brandying corrects the
acidity by precipitating the excess of cream of
tartar, to which the acidity was due.

" *Upon the action of alcohol on the free acids.*
With regard to free acids, alcohol, and as a
consequence, *brandying*, reacts upon them, but
in a more remarkable manner, for its effect is
two-fold, since it converts a substance essentially hurtful into substances most precious
and agreeable.

"Chemistry teaches that acids possess power
of combining with the essential principles of
brandy, and with them forming ethers. Now
those ethers, and particularly those formed by
the action of acids on wine alcohol in presence
of the other constituents of the wine, possess a
balsamic odor, and often a very energetic but
also delicate taste.

" Meanwhile, in order that the ethers may be
formed, even with length of time, at the ordinary temperature in a very weak acid wine, it is
necessary that it should contain a fair percentage of alcohol, and 10 per cent. is by no means
too much, when we wish to find some result at
the end of a few months, as should be the case
when we treat of moderate quality, which, to
save the expense of cellarage, should be drank
young.

"In order that the acids may become neutralized and ethers formed in a weak sour wine, it is absolutely necessary to fortify in the vat, so as to utilize a force, called by chemists '*the nascent state,*' and the high temperature developed during the process of fermentation, for warmth greatly favors this kind of reaction.

"Now we see in this instance the brandying has not primarily for its object to combat any injurious element, but to transform it into a precious product, which imparts to wine its most delightful relish. Facts known and borne out by ancient practice are in strict accordance with theory, though recent, but based upon syntheses the most complete, and analyses the most precise. The name of their eminent author, Berthelot, is no mean authority for their value.

"*On the influence of brandying on the color of red wines.*—Red wines owe their color to a coloring principle, which, with the exception of the '*tinta*' grapes, of which but few are cultivated, exists only in the skins. This coloring principle of dark red inclining towards blue is soluble in the alcohol, but insoluble in the other constituents of wine; consequently, limiting ourselves to this fact, the wines would be so much richer in color, in the relative abundance of spirit, in proportion to the coloring matter contained in the skins. Meanwhile, we know that there are wines of low alcoholic strength which possess a very beautiful color, but when we consider them we find them to be very acid, generally speaking.

" From this it seems to follow that acidity has a good deal to do in the coloring of wines, and such is the fact.

" In point of fact the red coloring principle is very easily oxidized under the united action of air and acid liquids, but in this instance the dark passes to a bright red, tending toward an orange tint, and from being insoluble in acid liquids it becomes soluble in them, while at the same time it ceases to be soluble in alcohol. Consequently with these new data, we can understand how, if the acid grapes be properly crushed before being put in the fermenting vessel, if they were again and again mixed up just previous to the setting in of fermentation, and above all, if they have a good long time on the skins, that is to say, exposed to the action of the air by every means possible, the wine will have a very deep color. Certainly there will be developed in it abundance of red coloring matter, which will be dissolved in the acid body of the wine, and the red coloring got in this way, being added to the purple extracted by the alcohol, will form a liquid with a color deeper, in proportion to the coloring matter in the skins. Such is the theory, and such the practice. '*This wine wants color; it never had sufficient cutting,*' say the vintners, and the vintners are right.

"From all this we may conclude *à priori*, that the addition of brandy, except brandying in the fermenting vat, would be useless, so far as coloring is concerned, unless for wines of low alcoholic strength and not much acid. Unhappily such is not the case, and under the consideration of the coloring, as of many others, they have much need of it.

"It is not in fact without risk, that we bring musts still warm from fermentation having set in, in contact with much air. Do what we will, and people are often heedless about it, the wine becomes more or less sour; it becomes in part vinegar; and in exchange for the beautiful color given to it artificially by coloring matters, it becomes the most disagreeable to the palate, and the most indigestible conceivable.

"Consequently for all light red wines, whether acid or not, brandying in the fermenting vat produces the happiest results upon their coloring matter. As concerns fortifying in the cask it is clear that as there are no skins for it to act upon, it can do nothing to assist the coloring; but if it does no good, it does no harm.

" We have said already that brandying pre-
cipitated the cream of tartar, and neutralized the
acids by forming ethers in combination with
them. Now. if we add brandy to wines, after
being put in cask, which owe their color in
great part to their acids, necessarily the red
coloring matter, which is not soluble except in
presence of the acid salts, will disappear and
fall down with the lees; and this is just what
happens. If any one doubts, let him ask the
vineyardists of the South (of France). Truly,
as it happens, this is a matter of minor consideration with them, as their wines are overload-
ed with color, and become rather an advantage
as it helps them to clear and brighten; but it is
a different matter in the case of deep-colored
acid red wines.

" Consequently, as far as color is concerned,
brandying, except brandying while in the fermenting vat, is useful for wines of low alcoholic
strength, whether acid or not—and the addition
of brandy to wines of the South (of France)
which are but slightly acid, while it would be
beneficial rather than otherwise to them, would
be highly injurious to acid wines of other districts; and consequently if some day free use of
brandy be allowed to all France, care will have
to be taken about this important difference.

" On the different kinds of fermentation, the

addition of brandy exerts even a greater influence on the keeping of wines, but this influence may prove either salutary or injurious, according as it is used with due precaution or not.

"A liquid like the must of the grape, left to itself at the ordinary temperature, if it does not heat much in the act of fermenting, ferments always spontaneously, and the fermentation is alcoholic, and nothing but alcoholic; it is the sugar which is split up —part forming alcohol, and the rest disappearing as gas. Should the temperature, however, rise above 30° centigrade, the sugar may still be broken up, and the result be lactic acid, in which case the wine must suffer severely. Happily. in the case of grapes, this kind of fermentation rarely happens, and so we wont waste time about it.

"Meanwhile, if at a temperature of 20° or 25° C., the air has free access, there is a tendency to form acetic acid, and to produce more or less vinegar. In this instance the vinegar is formed at the expense of the alcohol. By the heat of the cellar, however, as time goes on, with free access of air, the acetic fermentation may be produced—above all if the vinegar germ is already in the wine. This is not uncommon when the casks are left to lie long without having been filled up completely.

"As to the putrid fermentation, which is the

last, this is not caused by the saccharine principle of the grapes like the others, but from
some albumenoid principles, now but little
understood, perhaps from the ferment. This
disease is most frequently noticed in wines with
little acid, alcoholic or not. From all this we
conclude that the alcoholic fermentation makes
the wine, the other fermentations ruin it.

" *On adding brandy to wine in the cask, as
affecting its keeping properties.*—In the meantime
what are the reasons why the wines of the South
(of France), even the most spirituous, will not
keep without additional alcohol?

"It is necessary, first of all, to distinguish
between alcoholic wines and such as have hardly 10 per cent. of alcohol, because in the South
there are many of this sort, the old-fashioned
boiled wines. (*vinhos de caldeira.*)

"Confining ourselves for the present to these
latter, it may suffice to observe that they are at
once, both only slightly acid, moderately alcoholic, and deficient of tannin, that is, they
are short of the three elements which make wine
keep, and beyond this, being in the warmest
climate of France, they do not fulfill any of the
requisite conditions of a firm wine; they are not
that—and they easily change.

"Now, as to the very spirituous wines, the
cause of their want of firmness is somewhat

more complex. The grapes with which this class of wines is made are so saccharine that if all their sugar should be converted into alcohol, as is the case with moderately saccharine grapes, the strength would often exceed 18 per cent. of alcohol. Since fermentation stops at this strength, [*or at one a good deal lower*— Translator] it follows that such wines contain nearly always an excess of sugar unchanged, which, upon the least evaporation of the spirit, the slightest elevation of the temperature, or the smallest diminution of atmospheric pressure —in fine, at any moment they stand ready to return to fermenting. Well, among the thousands of these possible occurrences, more or less perceptible, some one may be expected to happen every day; and so those wines are in a perpetual state of feeble fermentation, which is enough to hold them in a continual bad state, and which renders them more liable to pass into acetic or even *putrid* fermentation, which happens, however slightly the other preservative elements diminish in quantity or potency.

"Now, how to obviate these serious accidents? As to the very spirituous wines, it is necessary *to add water* at the time of the vintage, to modify their tendency to become too alcoholic, so that their excess of sugar may be broken up during fermentation, or else, when

6

they are being drawn out of the fermenting vats into casks, to add so much alcohol as to raise them above the point at which fermentation is possible, and this is the occasion when their alcoholic strength is raised to 18 per cent. Now, as regards the weak wines of the South, it is necessary to add to them cream of tartar and tannin, or to dose them with brandy, and this is what is done.

"*What we can say of alcohol in wine.*—Over and above its direct action, the alcohol diminishes the acidity of acid wines, develops in them an agreeable taste and perfume, increases their color, and thereby renders useless the dangerous practices had recourse to for that purpose.

"Brandying produces all these felicitous results, and more than all, imparts to wines such a firmness that under its influence they can face the longest voyages under the extremes of temperature, and resist the deleterious action of the most unwholesome cellars."

So far Baron Thenard.

Reviewing all this exposition, it appears that, in the opinion of Baron Thenard, brandying is an operation useful and even necessary in many cases for the improvement and keeping of wines, and his views agree with the practice of many countries. Now, it is but due to him to say that he is the proprietor of very extensive vineyards in Burgundy, and that the rare wines of Burgundy are hardly ever under any circumstances fortified. The same is the case with the wines of the Medoc, and other vinicultural centres, which produce the finest of pure genuine wines. Fortifying may be useful and even necessary in many cases, but it is not indispensable for all wines, like racking and clarifying. If we read attentively the above extracts from Baron Thenard's works, we shall easily perceive that it is only in the treatment of wines full of acid, with little alcohol, and such as are flat and deficient in tannin and tartaric acid, although they have an excess of sugar and color, that the use of brandy becomes necessary. When the component principles of wine are nicely balanced, as is the case in ripe wines, made and kept in good condition, the addition of brandy is, to say the least, useless. Of course, it may operate as a remedial agent in wines by nature defective, such as, for example, most of the green wines of the Minho (Portugal), or even for mature wines naturally coarse and badly made.

In effect, alcohol is the most active preservative agent of wine, still when its percentage goes outside of certain limits, the wine loses those properties which are the most valuable as a hygienic and alimentary drink. All wine holding more than fifteen per cent. of absolute alcohol becomes an exciting beverage, needing to be used in moderation, and cannot enter into the list of good food wines. Wines of that strength, and upwards, have their place among liqueurs and fancy wines; and the modern taste relegates them exclusively to the department of liqueurs and luxuries.

Thus all the brandying that we practice on wines of food and daily consumption, should never go beyond fifteen per cent. of absolute alcohol, or rather 14.5. Binding ourselves by this rule we shall obtain the advantage of being able to introduce our wines into England under the one shilling (25 cts.) per Imperial gallon, just as the most part of the French wines are admitted, and will be so long as the present tariff exists.

Brandying may be done in various ways, and on different occasions. It may be done in the casks, or in the fermenting vessel before the wine is drawn off. In Portugal the brandying is in general done in the casks. Some do it just as the wine is put into cask, before the slow fermentation has ceased. This is what is usually done in the Douro, in making port wine; which in that particular instance may have a *raison d'etre*, which does not concern us now, but which finds no place in the making of pure genuine wines, because it would stop the light fermentation so necessary for the proper formation of wine. Others, and this is the case generally, add spirit only after racking in March. This brandy can be added only to increase the alcoholic strength, make up for any little loss of spirit by racking, or help in the clarifying which then takes place.

The use of brandy during fermentation proper is not in use in Portugal; still, writers on wine matters recommend it as the most convenient and efficacious way in cases where it would be either useful or necessary. And beyond all doubt if the object be to establish equilibrium among the elements of the wine, or to supply the want of sugar which should have supplied the spirit, or to diminish acidity and forward the formation of ethers, or to dissolve a larger proportion of coloring matters, on no occasion will the effect of it be more useful than at the cessation of the tumultuous fermentation; while yet the liquid is warm and in contact with the skins, etc. The quantity of brandy to be added in this case must be calculated upon the relative saccharine richness of the grapes, and always so as never to be excessive.

The brandy which should be used to fortify wine, under whatever circumstances this opera-

tion has to be performed, must be good wine
brandy distilled from wines of low alcoholic
strength.   Alcohols obtained from grains, from
beets and other roots, cane sugars, etc., rectified and afterwards reduced to any convenient
strength, how pure so ever they may be, and
perhaps on the very account of their purity,
never harmonize with wine in the same way
as good brandy directly distilled from wine.

In like manner, with alcohol chemically pure
diluted with water, *cognac* cannot be made, unless the proper spirit of wine be present,
(*onanthic ether*, *etc* ,) for all that its chemical
composition may be identical.   What practice
has made clear, is that the purer and more
highly rectified the alcohol is, the more reluc-
tant is it to amalgamate with the component
principles of wine of a delicate constitution.

The need of brandying afterwards may be
anticipated and supplied by the addition of
cane sugar while fermentation is in progress.
By being fermented, this sugar furnishes the
requisite amount of alcohol to fortify the wine.
The price of the very finest sugar (the only
sugar which can be used without introducing a
bad taste into the wine), is the chief reason for
not using it for fortifying weak acid wines, but
when we speak of weak acid wines, like the
"green wines" of the Minho, which have not
more than 6 or 7 per cent. of alcohol, the ben-

efit in relation to both strength and quality de-
rivable from this practice is manifest.

I cannot here and now refer to any results of
direct experiments made in the Minho with the
above object, though I have recommended it
more than once. However, this plan of improving and fortifying weak acid wines, has
been in use in other countries for a long time,
and it would be well if our viniculturists of the
Minho would give it a trial.

According to the practice adopted in Burgundy, the sugar should be put in when fermentation is nearly over, the quantity being
1,700 grams (3 pounds 8 ounces) for *each degree* of strength which we wish to produce in
each hectolitre of wine.

The sugar must be first dissolved in a sufficient quantity of the same must into which it
is going to be put, and when the solution is
added to the fermenting liquid, the whole must
be thoroughly stirred and incorporated. Upon
the addition being made, fermentation sets in
anew, and as soon as it ceases, the wine may
be drawn off in the usual way.

Some practical viniculturists have observed
that wines so treated, show for a long while an
inclination to ferment and to turn, but it does
not always happen; but to prevent any danger,
the means are at hand. It needs only that we
add 2 litres of brandy to each *hectolitre* of wine

before drawing it off. Common pomáce brandy serves for the purpose; for according to Viscount Vergette—Lamotte, the offensive taste of such brandy is not imparted to the wine when put into the fermenting vat. The cost of this is quite trifling, and that of the sugar not very much. In the instance of wines which have to be sold very cheap, the additional cost will be felt; but it is better for the purchaser to obtain a decent article, and for the proprietor to have a firm wine than a detestable beverage which does not deserve the name of wine. On this account I do not hesitate to advise the vignerons of the Minho who may try to improve their green wines, to do so either by adding sugar during fermentation, or by adding brandy *before* drawing off—for that is what they need, and not brandying after being put in cask.

For the same purpose as sugar, there is commonly used in Portugal and other countries, fresh must, concentrated to the consistence of syrup, called *arrobe*. It is then ready to be added to the fermenting must. Where the musts and wines are very acid, as is the case in the Minho, often containing less than 10 per cent. of alcohol, arrobe would be out of place, because the acids and salts are fixed in it, and would only augment the quantity already too great in the wine.

METHOD DEVISED BY M. LOUIS BARRAL.

I will now explain a process devised by M. Louis Barral, a scientific chemist and vineyard proprietor in the Herault, which is destined to improve, and cause ordinary wines, whatever be their nature, to keep good. At the International Exhibition of 1867, a medal was awarded to it on account of the excellent results obtained by the use of it. The process is very easy, and within reach of any vineyardist, as may be seen in the condensed form, as given by the inventor himself. In his condensed form we have an epitome of every operation of wine-making; and so those who think well to adopt his invention, may follow his instructions literally.

Following is the regular order of procedure:

1st.   Tread, crush, or by any means reduce the grapes to a uniform pulp.

2d.   Leave the must to ferment for only five or six days under ordinary conditions of temperature, say 16 to 18 degrees Centigrade.

3.   Draw off all the wine and put it in casks apart.

4.   Press the pomace.

5.   Take from a portion of this pomace while quite fresh, the skins and seeds, and a portion of the stalks.

6.   Introduce this selected portion of pomace into a cask with a large bung; then pour over it as much high proof brandy as will cover it, and leave it to macerate, after hermetically closing the bung hole, and never disturbing it until February following, when it is necessary to rack it off.   This is what M. Barral calls *"tannic wine alcohol."*   To get the residue out of the skins, etc., it is necessary to wash them with two or three successive portions of wine, and finally to press them.   The wine and pressings must be kept separate from the brandy.   When they have become clear they need racking.

To improve ordinary wine, which was drawn from the fermenting vat into casks, it is necessary to employ one litre and a quarter, or a litre and a half of the "tannic wine alcohol" to each hectolitre (about 3 pints to 26 gallons). The wine got by washing the pomace may be used at the same time, and in the same proportion.   When the wine has been prepared as directed, it is desirable to clarify it, and after a few weeks, rack it.

The process just described and employed by M. Barral on ordinary wines of bad years in

the south of France, is equally applicable to wines which are always ordinary in many countries. The foundation of it is the addition of brandy charged with tannin, one of the best preservative principles of wine. The clarifying with isinglass or gelatine (*colla*), which should always follow the brandying, will soften down any asperity or roughness which possibly may have been communicated by the tannin. By these operations the wine gains in strength, in keeping power and in color.

In districts where on account of the generally good quality of the wines, the above process becomes needless; nevertheless, it is well to prepare a certain quantity, both with the red and white grape pomace, separately, for it may be required, either to facilitate clarifying when there is a deficiency of tannin, or to treat wines which become diseased for want of it.

ON WHITE AND PALE WINES.

(VINHOS BRANCOS E PALHETES.)

Desiring as I do to limit this treatise to an exposition of the most convenient methods of preparing pure red wines of general consumption, I might have properly omitted all mention of *white* wines, the manufacture of which among us is very limited, relatively to the red. As to the pale wines (*palhetes*), the making of them is perhaps a little more extensive than of ordinary white wines. Still they are made for local consumption only, and trade, especially for export, does not purchase them. I consider, however, that I ought to say something about them, because, though the business in them is small and local, there is no reason why they should not have a position in foreign markets, and sooner or later a good business may be done in them.

Setting aside especial kinds of white wines, chiefly liqueur wines, such as the Muscatels, the Malvasias, and the wines of Madeira, of which I do not intend to treat in this treatise, one may say in a general way that the white wines of Portugal belong to two distinct classes, viz., dry white wines for ordinary consumption, and alcoholic ones, the chief use of which is to blend with red wines, to give them the semblance of age. In all the essentials of wine-making there is no difference between red and white wines. They are made exclusively with white grapes, whilst the reds are made either with red grapes alone, or, as is more usual, with a mixture of white and red, the largest part being red. The pale or pallet wines are made with a preponderance of white over red (Schiller wines); consequently there is a wide difference between our method of making white and *Schiller* wines and those in vogue in foreign wine countries.

Full-bodied, spirituous white wine, principally those used in the commerce of Oporto to mix with red wines, are made exclusively with white grapes, and in exactly the same way as the red; and consequently, since they are fermented on the skins and stalks, they acquire an exceedingly deep yellow color, partly, no doubt, from the oxidation of the organic matter of the skins. They receive a very considerable proportion of brandy immediately after being drawn off the fermenting cask, which preserves a certain amount of sweetness and preventing further fermentation. Wines of this character, very valuable for the purpose for which they are intended, can in no sense be considered as genuine food wines for ordinary consumption; therefore I consider this is not the place to treat of them.

The making of dry white wines differs from the foregoing only in so far as the slow fermentation in the cask is not suspended nor the second fermentation by the use of brandy, but allowed to become as dry as possible, the same as red wines. From the nature of the grapes of which they are made, white wines deposit a much more considerable quantity of lees than the reds, and consequently require more frequent racking, and at least one clarifying with isinglass, or where it cannot be had, with the best gelatine, aided by an addition of cream of tartar.

Musts of white grapes fermented on the skins like those of red grapes, yield wines equally dry, relishing and wholesome, and may take their place among food wines, and are preferred by some to red wines, as being lighter; still, the great majority of men prefer red wines for every-day use. It is not meant here to controvert what statistics show plainly enough concerning the consumption of white wines in England, which is far greater than that of reds; but we should not forget that the greater part of the white wines drank in England are called

Sherries, generous alcoholic wines, which, just like the Ports, belong to an entirely different class. If we compare only the genuine red wines (the French Clarets, for example) and the French whites consumed there, and excluding the Champagnes and liqueur wines, we shall perceive that the difference is considerable in favor of reds for daily use. Whatever the cause of this preference may be, the fact is that the actual increase of consumption is altogether in favor of pure red wines, without excluding, however, whites of a like nature, still requiring certain qualities to be present in them which cannot be imparted by our present Portuguese methods.

Genuine white wines must be free from all roughness and astringency, must have a slight sweetness and be very light, perfectly clear and bright, as well as nearly colorless. These qualities can be secured solely by fermenting without the skins or stalks; and so the methods in general use, both in France and Germany, with very slight modifications, consists in pressing the grapes and allowing the must to run, and putting it at once to ferment and leaving it to run through its whole course, and giving it afterwards the same care and cleanliness as required in the case of red wine. The statement of the process is simple enough,

but the cares and precautions required in practice are very minute when the safest and happiest results are aimed at. Above and before
all, there is indispensable need of the most extreme cleanliness of all utensils whatever, that
are to be brought into contact with the grapes
or with the must, or with the wine when made.
If the grapes are to be trodden before being
pressed, which is convenient, for as much as it
can be rapidly done, allowing the pulp to run
out so as not to allow fermentation to set in,
the bench and press ought to have been scrupulously washed even till the water ran out
clear, to obviate the introduction of any foreign matter which by possibility might impart
its own taste or smell to the wine. The same
cleanliness is demanded for the casks, etc.,
which should never have been used for any but
white wine, and supposing them to be new, any
taste of the wood should have been washed out
of them. For white wine the grapes should
be quite fully ripe, and all dry, imperfect or al-
tered berries should be picked off them.

Generally, the must of white grapes deposits
a deal more lees than that of red grapes, and
as a consequence it clears with far more difficulty and is liable to repeated fermentations.
In order to avoid this trouble, and preserve the
sweetness which it ought to have, and that fine

texture which forms one of its most highly es-
teemed qualities, it is desirable to remove as
far as may be, the bulk of the sediment, even
before fermentation becomes established. Two
methods are in use to do this: 1st. Some
makers receive the must in comparatively
small tubs with taps, and as soon as the must
has formed a thick scum on the surface they
decant through the tap the wine now separated
from the sediment that has fallen, and from the
scum which has risen to the surface. This
sort of decanting is carried on with other tubs
like the first, so long as the must goes on forming new sediments and new scums, in fact till
it ceases to form thick scum, when it is put
into casks, and when the fermentation will form
and work through more gently than had the
above operations not taken place. The casks
should be filled quite full, so that the scum
that forms during the more energetic stages of
fermentation may escape freely through the
bung-hole, and not fall into the body of the
wine where they would be ready to renew fermentation. When fermentation is quite finished, the casks should be filled up with the
same kind of wine, always, and subjected
to all the same care and precautions as
mentioned already for red wines. 2d. The
other method of preserving a part of their
sugar in white wines, which is easier in prac-

tice and equally effective, consists in putting
the must, just as it is made, in casks well sulphured, in order to arrest fermentation for
some time.   Under these conditions the greater
part of the gross sediment becomes deposited
before fermentation commences, and as soon
as the must looks milky it is racked or trans-
ferred to other casks similarly sulphured.
This operation may be repeated as soon as fermentation begins to appear, and again, if needful, until the greater part of the thick sediment
has been removed.   The fermentation is in
this way retarded by the sulphurous gas, and
the larger portion of the ferment is eliminated,
and so a certain portion of the sugar is preserved
in the wine and the taste of the fruit, so pleasant to the palate.

Fine, sweet white wines need to be clarified
as completely as possible, to enable them to
keep.   This is an indispensable condition.
Isinglass, in the proportion of 50 grams
(1½ pounds) per pipe.   This isinglass should
be first well bruised with a wooden mallet,
then cut very fine and put to soak in a small
quantity of the wine we are about to clarify,
and which is to be repeated until the isinglass
will absorb no more; then a little hot water is
to be added and the whole rubbed up well and
strained through a cloth.   To the thick liquid
we add a quart or so of wine and beat it well

once more, so as to raise a froth, and then use it in the ordinary way. When the wine has been clarified, it is good practice to always add a small quantity of good brandy.

White wines can be made with red grapes, and good writers on wines consider them the best and finest; still the bulk of the white wines are made with white grapes, unless the champagnes, in which red or white are used indiscriminately.

The wines (*vinhos palhetes*) which are made expressly with brief fermentation on the skins of the red grapes, are very pleasant, because they are light, soft and delicate. They are intermediate between the reds and the whites. The making of them presents no difficulty.

After making red wine, if the skins show no sign of alteration, they can be utilized along with the white grapes to make an excellent Schiller wine, of a more or less deep color, in proportion to the length of the fermentation on the red skins. In this way the wine is soon made, being forwarded by the state in which the red skins are when used; but under all circumstances the wine should be got into casks before fermentation entirely ceases, to prevent its acquiring the harsh taste of the pomace. The Schillers ought to be light and soft, and as a consequence, should mature and become ready to be used in a short space of time.

BOTTLING.

The proprietor usually sells his wine in cask, and merchants for the most part export it in the same manner; nevertheless there is no doubt that if the grower had the means of keeping it till it was fit for consumption, or for bottling, and he could with economy do it in his own cellar, he would have among other considerations the satisfaction of knowing that its reputation rested on its intrinsic merits. The producers of fine wines have a great advantage in selling them in bottle and with their distinctive labels. On this account, I think it worth while to say something about bottling, limited as my remarks necessarily must be in a compendium of wine matters, like the present.

Bottling is an operation needing to be carefully performed, for in many cases the achievement of the highest qualities of the wine depends on it. Generally speaking, fine, delicate wines do not arrive at their maximum of perfection till they have been for some time in bottle.

Many are the conditions needful to be attended to, in order to do this last work in a proper manner, such as the state of the wine, and of the weather, the materials employed, and the method of working.

As to the state of the wine, it must have attained such a condition of ripeness as that it can get no more good by being left longer in cask, a point which varies not only between one wine and another, but between the same wines; that is to say, between wines grown on the same vines, and on the same soil, but of different vintages, and even between casks of identical wines, for it is well known that not only do the wines of some years mature sooner than those of others, but those of the same year, of the same vintage, and of identical make, will vary, due only to their having been put in separate casks, and subjected to slight differences of temperature. There is no better means to judge of this than the palate of a cellar man accustomed to treat them. Wine must never be bottled unless it be perfectly still, clean and transparent. The smallest want of any of these conditions might easily ruin it, consequently it becomes indispensable to clarify before bottling.

From what has just been said, it is clear that no particular age can be assigned for bottling wines. In all cases, it ought neither to be so old as to have begun to decline, nor so young as not to have shown its predominant qualities. Of the two extremes, the better would be to bottle while it is too young. Inside the bottle, where no air can get to it, it exists under different conditions from what it had when in the cask, where its vapors could escape and air could enter. Nevertheless, reactions continue to go on among its constituent principles, the results being new products, as yet but little studied, which improve its quality and develop its bouquet—consequently it ought to have abundant vitality when bottled.

The best time for bottling is perfectly calm and still weather, during autumn and winter, as far as the end of March, when the atmosphere is serene and cold and dry, with no south wind nor threat of thunder, is a good time to bottle wines. The reason is the same as that for racking under similar circumstances.

Attention must be paid to the quality of the bottles and the corks. Recent observations have proved that the quality of the glass of which bottles are made is not a matter of indifference, as concerns the preservation of the wine. The glass of which ordinary bottles are made is a multiple silicate of alkaline bases, both earthy and metallic; all the more easily fused and wrought in proportion as the alkaline bases used are more abundant. Now the softer the glass is, the more readily it is attacked by the water and acids contained in wine; for the excess of alkali may become dissolved in the water, or more likely be combined with the acids, to the injury of the composition of the wine. In such bottles, green wines holding much acid, cannot be kept long; and even the others, rich as they may be, run the risk of injury. Consequently much care is necessary, and as few can verify for themselves, whether the glass is good or not, the safest course is to ask an expert in such matters.

When satisfied that the quality of the bottles is what it should be, the next indispensable step is to have them washed both inside and out, perfectly clean, till not a speck remains, nor any foreign body, or any smell whatever. Never, on any pretext use shot. There is always great danger of some corns remaining fast between the bottom and sides of a bottle, and escaping notice, which will afterwards become dissolved in the wine and impart poisonous properties to it, and not alone from the lead, but also from the arsenic which usually is to be found in shot. Metallic chains made of small rings on purpose, are certainly preferable for loosening impurities in old bottles.

When the bottles have been washed as described, they must be left to drain till they are perfectly dry; and for this purpose must be placed sufficiently long with the neck down in some kind of frame or rack.

For closing bottles, hardly anything but corks are used, and it is very essential that the cork be of good quality, fine, solid and elastic, and the new corks well cut and free from defects.   Bottling presents no difficulties, but needs discernment, cleanliness and practical experience.   For the rest, the barrel about to bottled should have been placed at some sufficient height to allow the wine to be conveniently drawn off, and as it is customary to remove the bung it is always well to fix over the hole some cotton wool or lint, to filter the air and prevent any germs from entering that might injure the wine.   The same precaution should always be taken when wine is used from the cask, and even still more carefully, on account of its longer exposure to the air.   In place of the simple cotton filter it is safer to fix in the bunghole of the cask out of which wine is drawn for daily use, a tin tube in shape of the letter S in the curved part of which the cotton wool is to be put (or water) through which the air has to pass; and while doing so, free itself from germs and dust, which it brought along with it.

The bottles ought to be filled completely, and when possible should be corked with a corking machine, and needle to admit the escape of air. When however that cannot be done, the next effort should be directed to leaving the least possible air-space between the bottom of the cork and the top of the liquor. The corks should be well soaked either in the wine to be bottled or in good brandy. These bottling machines are quite common now-a-days, cost little, are very handy, and save time and labor when considerable work has to be done, and prevent breakage as far as possible.

There is a vast advantage in the cork's touching the top of the wine, to the entire exclusion of air. This is obtained by the use of what is called the *"needle."* It is a small steel instrument with a slot, which when placed against the neck of the bottle, forms a channel to admit the escape of all air. If the bottle be full, the pressure of the cork upon the fluid will force as much out as is required. When the cork is driven home, the needle is drawn out, and the space it occupied becomes filled completely by the cork.

Supposing the bottles to have been well corked with good carefully made corks, the bottles may remain for a long time without need of either capsules or sealing, as is usually done when they pass off into commerce, provided the cellar is dry and airy. Unfortunately, the greater part of the corks met with in trade are far from being good, so much so as to render it impossible to brand them with hot iron, as is

7

customary in the best wine countries; and in
this case it is indispensable to cover the top of
the bottle and cork with a composition of rosin,
Burgundy pitch, a little yellow wax and some
coloring matter, such as red lead.

After bottling, and while the wine is under-
going its changes, it ought to lie at rest for a
good long time, if not a whole year. During
this period it is proper that the bottles should
lie in a clean store, and always on their sides to
keep the corks moist. Light table wines of low
alcoholic strength are very liable to take
"flour" if the above precautions of excluding
air at the corking be not attended to. If the
wine after bottling forms a deposit, as often
happens, it will fall to the low side, and in
handling the bottle, drawing the cork, and decanting the wine, nice care should be used.

From the moment the wine has been bottled
with all the precautions above indicated, we
might suppose it to be out of the reach of the
air's action, and any effect it could produce; still
changes go on in it, which for the most part
improve it, slow though they may be. These
changes are liable to be affected by temperature
of the cellar, by light, and by being shaken; so
the cellar itself requires to be considered. Temperature and light are essential conditions.
Wines that are not very alcoholic gain by being

kept in a temperature at about 10°·Centigrade, or 50° Fahrenheit, and where the light is diffused and not very bright. Direct sun-light is always injurious to such wines, not alone by unequal heating, but by its action on the coloring matter. Full-bodied wines of much alcoholic strength improve best in stores whose temperature is much higher than that for table wines. In a future chapter concerning the improvement of wines, I shall have occasion to treat at some length on the action of heat on wines.

Having now followed the process of wine making, as far as pure, genuine, alimentary wines are concerned, from the vintage to the bottling, when they undergo their last and final stage of perfectioning, I now consider the first part of this work finished, not without the conviction of having said all that is necessary to induce and animate our vineyardists to manufacture those kinds of wine best suited to increase a demand for our produce, and throw life into our commerce. I have by no means exhausted the subject. It needs still large development and special teachings, which may at a future day form the object of a more extensive treatise; but in writing this book, *my supreme and only intention is to place in the hands of our viticulturists compendious instructions,* to enable them to follow along the way which, in my conscience, I judge conducible to the revival (regeneration) of our viticulture, and more chiefly still to that of the commerce in our wines.

www.ingramcontent.com/pod-product-compliance
Lightning Source LLC
Chambersburg PA
CBHW030600270326
41927CB00007B/990